MONTGOMERY COLLEGE LIBRARY
ROCKVILLE CAMPUS

The Musculoskeletal System

STRUCTURE AND FUNCTION IN DISEASE
MONOGRAPH SERIES

ABNER GOLDEN, M.D.
Series Editor

Available Volumes
Golden & Maher: THE KIDNEY
Robertson & Dinsdale: THE NERVOUS SYSTEM
Kashgarian & Burrow: THE ENDOCRINE GLANDS

**STRUCTURE AND FUNCTION
IN DISEASE
MONOGRAPH SERIES**

The Musculoskeletal System

LEON SOKOLOFF, M.D.

Professor of Pathology
State University of New York at Stony Brook

JOHN H. BLAND, M.D.

Associate Professor of Medicine
University of Vermont

 The Williams and Wilkins Company • *Baltimore 1975*

Copyright ©, 1975
The Williams & Wilkins Company
428 E. Preston Street
Baltimore, Md. 21202, U.S.A.

All rights reserved. This book is protected by copyright. No part of this book may be reproduced in any form or by any means, including photocopying, or utilized by any information storage and retrieval system without written permission from the copyright owner.

Made in the United States of America

Library of Congress Cataloging in Publication Data

Sokoloff, Leon, 1919–
   The musculoskeletal system.

(Structure and function in disease monograph series)
Bibliography: p.
   1. Musculoskeletal system—Diseases. 2. Musculoskeletal system. I. Bland, John Hardesty, 1917– II. Title. [DNLM: 1. Musculoskeletal system. 2. Musculoskeletal system—Physiopathology. WE100 S683m]
RC925.S64   616.7   74-30473
ISBN 0-683-07888-7

Composed and printed at the
Waverly Press, Inc.
Mt. Royal and Guilford Aves.
Baltimore, Md. 21202, U.S.A.

To Bev and Libits

FOREWORD

The Musculoskeletal System by Sokoloff and Bland is the fourth member of the *Structure and Function in Disease Monograph Series*. It is again the combined effort of a pathologist and an internist, following the format of the earlier volumes.

Leon Sokoloff recently returned to academia after 20 years in the Public Health Service. During his tenure as Chief of the Section on Rheumatic Diseases, Laboratory of Experimental Pathology, National Institute of Arthritis and Metabolic Diseases in Bethesda, he had extensive experience with descriptive and experimental pathology of arthritis and rheumatism. His previous book, *The Biology of Degenerative Joint Disease*, is a widely cited and authoritative analysis of the subject.

John Bland was Director of the Rheumatism Research Unit, University of Vermont College of Medicine, from 1950 to 1970. His earlier books have dealt with a number of subjects: lichens and mosses, rheumatism, and, his first love, salt and water metabolism. From the latter it was a natural step to the connective tissues. His final turn to the rheumatic diseases also followed logically and was inspired by the late Walter Bauer. The role from which Dr. Bland derives deepest professional satisfaction is as clinician and teacher.

Drs. Sokoloff and Bland share wide-ranging interests and have enjoyed a warm friendship for many years. It was during Dr. Bland's sabbatical year in Bethesda that the strategy for the present volume was worked out. That their mutual respect and affection has survived completion of the work is one measure of their friendship.

ABNER GOLDEN, M.D.
Series Editor

PREFACE

The stated goal of the *Monograph Series* is to offer an integrated presentation of structure and function to all students of disease. This covers a lot of territory, and we therefore wish to spell out what we are trying to accomplish in this volume.

We find it difficult, first of all, to come to grips with the concepts of structure and function. It would seem more appropriate to a metaphysical disquisition than a medical textbook to distinguish between them when every day brings discoveries of life processes that can only be understood in terms of subcellular and molecular anatomy. Structure and function are only different ways of looking at the same reality. The relevance to the present work is that the integration at issue is the team teaching of what has traditionally been taught separately by pathologists and internists. Professional bodies of pathologists have wrestled for the past two decades with the problem of what constitutes pathology. In a generic sense it is the study of the cause and nature of disease, but this concept is so broad as to be meaningless without some defining operational characteristic. We are not offended by the idea that pathology still corresponds to pathologic anatomy. It basically describes biologic processes in terms of what can be made visible in tissues at points in time. The ultimate worthwhileness of team teaching has yet to be determined. Although the amount of time and effort demanded of faculty to coordinate such teaching may ultimately prove prohibitive, we think there is much good in it and have prepared the text with this in mind.

The subject matter treated is more restricted than the prospectus might indicate. This is not primarily meant to be a textbook of clinical medicine but to teach the biologic concepts of disease in a clinical perspective. The choice of subjects covered necessarily is arbitrary. Those selected are there either because they are important in terms of public health or, if rare, because they illustrate a useful principle. Like some other volumes in this series, we have not dealt seriously with neoplastic or surgical disorders. Primary musculoskeletal tumors are relatively rare and difficult even for the specialist. An excellent source of information on bone tumors is included in the bibliography. Although the musculoskeletal system functions as an integrated whole, it comprises three distinct organ systems—skeletal muscle, bone, and joints. The first four chapters address the normal aspects of these organs; and those that follow, the abnormalities. In keeping with the orientation toward solving clinical problems, we have attempted to group the material about dominant clinical themes rather than disease entities. Accordingly many x-ray films are presented because that is the way skeletal structure is dealt with in practice. It is our belief that students should acquire competence in interpretation of such films early on. Beginners are unnecessarily intimidated by roentgenograms and these illustrations have been labeled in an attempt to help them locate the lesions.

This book is meant for medical students and, to a lesser extent, younger house staff officers and clinical fellows. It is not directed toward all students of disease. We are all students of disease. In places, overly simplistic concepts have been retained for their didactic usefulness although we are cognizant of recent advances that cast some doubt on their validity. Aside from its more catholic approach, the subject matter is treated in somewhat greater depth than in most standard textbooks of pathology. Nevertheless it is in no way an advanced treatise. We would feel well rewarded if our book served to introduce the reader to the more specialized monographs listed below.

We are happy to acknowledge the assistance of many people in preparing the volume. One of us (J. H. B.) was the recipient of a sabbatical grant from Mrs. Sarah Given Larson, the Irene Heinz Given and John LaPorte Given Foundation. Most of the illustrations were taken from our files but we also thank our colleague, Dr. **Aubrey J. Hough**, as well as Peter McCarthy and Eugene McDermott for preparing additional prints. Gary J. Nelson and Kathleen Gebhart made the drawings. Permission to reproduce material Leon Sokoloff has published elsewhere was granted as follows: *Figure 1.5: In* R. M. Kenedi (ed.): Perspectives in Biomedical **Engineering**. London, Macmillan, 1973, p. 136. *Figures 11.2, 11.4, 11.5, 11.6, and 12.1: In* J. L. Hollander and D. J. McCarty (Eds.): Arthritis and Allied Conditions, 8th ed. Philadelphia, Lea and Febiger, 1972. *Figure 12.3:* Ann. N.Y. Acad. Sci., 240:285, 1974. Permission to reproduce John Bland's illustration, *Figure 13.2*, was granted by the Journal of Rheumatology (1:319, 1974).

Drs. Lauren V. Ackerman, Norman R. Alpert, Mark D. Aronson, Edward S. Horton, Arthur J. Kunin, Huntington Mavor, Charles A. Phillips, and James B. Wyngaarden reviewed portions of the manuscripts critically. Dr. Mavor graciously provided the electromyograms from which Figures 4.6 and 4.7 were constructed. The various drafts of the manuscript were typed by Ray Sokoloff, Sylvia G. Vaughn, Jean Isaacowitz and others.

To all these we are most grateful.

LEON SOKOLOFF
JOHN H. BLAND

GENERAL REFERENCES

Bourne, G. H. (Ed.) The Biochemistry and Physiology of Bone, 2nd ed. New York, Academic Press, 1971.

Bourne, G. H. (Ed.) The Structure and Function of Muscle, 2nd ed. New York, Academic Press, 1972.

Copeman, W. S. C. (Ed.) Textbook of the Rheumatic Diseases, 4th ed. Edinburgh and London, E. & S. Livingstone, 1969.

Duthie, R. B., and Ferguson, A. B., Jr. Mercer's Orthopedic Surgery, 7th ed. Baltimore, Williams & Wilkins, 1973.

Hancox, N. M. Biology of Bone. Cambridge University Press, 1972.

Hollander, J. L., and McCarty, D. J., Jr. (Eds.) Arthritis and Allied Conditions, 8th ed. Philadelphia, Lea & Febiger, 1972.

Jaffe, H. L. Metabolic, Degenerative and Inflammatory Diseases of Bones and Joints. Philadelphia, Lea & Febiger, 1972.

McKusick, V. A. Heritable Disorders of Connective Tissue, 4th ed. St. Louis, C. V. Mosby, 1972.

Pearson, C. M., and Mostofi, F. K. (Eds.) The Striated Muscle. Baltimore, Williams & Wilkins, 1973.

Rasmussen, H., and Bordier, P. The Physiological and Cellular Basis of Metabolic Bone Disease. Baltimore, Williams & Wilkins, 1974.

Schmorl, G., and Junghanns, H. The Human Spine in Health and Disease, 2nd American ed. New York, Grune & Stratton, 1971.

Schubert, M., and Hamerman, D. A Primer of Connective Tissue Chemistry. Philadelphia, Lea & Febiger, 1968.

Sokoloff, L. The Biology of Degenerative Joint Disease. Chicago, University of Chicago Press, 1969.

Spjut, H. J., Dorfman, H. D., Fechner, R. E., and Ackerman, L. V. Tumors of Bone and Cartilage, Second series. Washington, D.C., Armed Forces Institute of Pathology, 1971.

Walton, J. N. (Ed.) Disorders of Voluntary Muscle, 3rd ed. Baltimore, Williams & Wilkins, 1974.

CONTENTS

1. CONNECTIVE TISSUES AS MECHANICAL AND LIVING SYSTEMS 1
   Mechanical function and properties. Chemistry of connective tissue matrix. Microscopic appearance of connective tissue matrices. Water and salt metabolism. The cells of connective tissues.

2. BONE AS TISSUE AND ORGAN 20
   Growth and development. Bone biodynamics. Vascular and nerve supply. Mechanism of mineralization. Regulation of mineralization. Effects of other hormones on bone. Repair of bone. Extraosseous manifestations of disturbed bone metabolism.

3. JOINTS 35
   Diarthroses. Amphiarthroses.

4. SKELETAL MUSCLE 43
   Muscle cells. Bioenergetics of muscle. The motor nerve fiber. The neuromuscular junction. The muscle spindle. Repair of muscle. Electromyography. Muscle biopsy.

5. THE PAINFUL MUSCLE 58
   Cramps. Inflammatory diseases of muscle. Miscellaneous non-inflammatory conditions.

6. THE WEAK MUSCLE 66
   Atrophy. Motor neuropathies. Muscular dystrophies. Myasthenia gravis. Miscellaneous myopathies.

7. DISTURBED SKELETAL GROWTH 73
   Short stature. Increased stature.

8. OSTEOPENIAS ... 81
Osteogenesis imperfecta. Osteoporosis. Osteomalacia. Osteitis fibrosa. Osteopenias secondary to neoplastic and other medullary lesions.

9. HYPEROSTOTIC DISORDERS ... 97
Paget's disease of bone. Osteopetrosis. Hypertrophic osteoarthropathy. Myelosclerosis. Caffey's disease. Osteofluorosis.

10. INFECTIONS OF BONES AND JOINTS ... 105
Osteomyelitis. Miscellaneous bone infections. Infectious arthritis.

11. NON-INFECTIOUS ARTHRITIS ... 114
Rheumatoid arthritis. The "variants" of rheumatoid arthritis. The collagen diseases.

12. NON-INFLAMMATORY JOINT DISEASES ... 128
Osteoarthritis. Special forms of degenerative joint disease. Crystal deposition arthropathies. Aseptic necrosis of bone.

13. BACKACHE ... 145
Structure and function of the spine. Ankylosing spondylitis. Degenerative diseases of the spine.

14. LABORATORY DATA IN MUSCULOSKELETAL DISEASES ... 152
Serum enzymes. Serum electrolytes and small molecules. Urine. Serologic abnormalities. Synovial fluids.

1

CONNECTIVE TISSUES AS MECHANICAL AND LIVING SYSTEMS

Bone and the articular apparatus share major chemical and biologic features with other connective tissues. All provide "support," *i.e.*, mechanical strength and compliance (Fig. 1.1) in proper proportion, to the parts of the body. Other functions, such as the selective transport of water and solutes from the blood stream to the cells, are themselves physical phenomena. These tissues necessarily are subject to the same principles of engineering design as non-living substances. They constitute what in materials science are called composites. Composite materials consist of at least two physical phases. Each phase possesses its own distinct properties but in its intimate interaction with the other imparts a different net effect than the sum of the two parts separately. In composite materials, one component characteristically is a fiber having tensile strength, *i.e.*, one which resists tearing when stretched. The other component is a weaker interstitial filler. Common to the various connective tissues is the fibrous phase, collagen, and the interfibrillar, water-rich proteinpolysaccharide complexes which collectively make up the ground substance.

MECHANICAL FUNCTION AND PROPERTIES

One would anticipate large quantities of collagen in connective tissues subject to high tensile forces. Furthermore the collagen fibers ought to be arranged along the lines of tensile force. Indeed, this is the case. In tendons and ligaments, the fibers are oriented in fairly parallel array in the axis of the tissues. In cartilage and bone, the arrangement is more complex, thereby giving insight into the mechanical function of these structures. Collagen is inextensible. Flexibility is achieved only by buckling of the fiber or by patterns in which it is "woven" into the fabric of the tissues. In this respect it differs from another fibrous protein, elastin, which is the principal component of the ligamenta flava and nuchae of the vertebral column. Elastin stretches 20 to 30 percent before

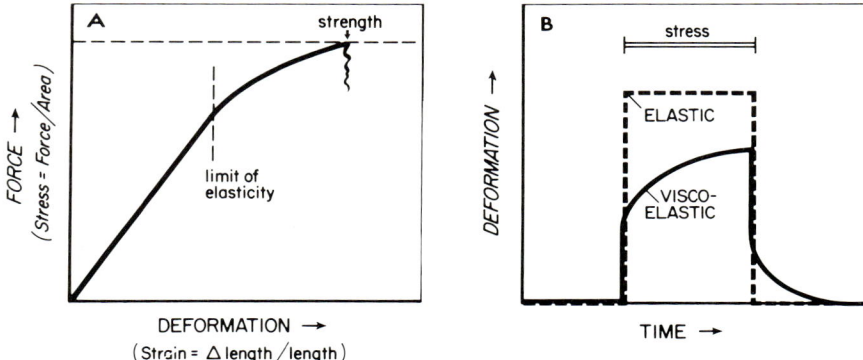

Figure 1.1. Physical responses of tissues to static mechanical forces (pushes or pulls). *A.* When the deformation is not affected by the duration of the applied force and is directly proportional to the stress, the tissue is elastic. The slope of stress/strain is the Young's modulus of *elasticity* and is a measure of the stiffness of the tissue. It is not equivalent to stretchability as in a rubber band. For bone the value is in the order of 10^6 pounds per square inch, while for cartilage, approximately 10^2. *Compliance* is the inverse of the elasticity. When the limit of elasticity is exceeded, the tissue is weaker and becomes deformed more easily. The stress at which fracture occurs (*broken line*) is the breaking *strength*. *B.* Connective tissues are more *viscoelastic* than truly elastic materials: they become increasingly deformed as time elapses. The progression of the deformation results from slow displacement of fluid in their matrix. The deformation at any given level of stress is much less in viscoelastic materials.

it tears, while the figure for collagen is only 1 percent or so. Although collagen is strong in tension, it necessarily has little strength in compression; fibers, by nature of their high axial (length:breadth) ratio, buckle when their ends are pushed together.

Connective tissues, therefore, gain their compressive strength from the water which constitutes the bulk of the ground substance. Water is incompressible and is restrained from flowing out of the connective tissue by the proteinpolysaccharides (proteoglycans), molecules so large and so constituted that they cannot escape from the fibrous mesh in which they are entrapped. These macromolecules are polyelectrolytes, often containing large, fixed anionic charges arising from their sulfate content. The electric charges are responsible for major physicochemical as well as histochemical characteristics of the tissues. They attract cations and are a major pool of electrolyte metabolism in the body fluids. Mutual electrostatic repulsion causes varying elongation of the ground substance polysaccharide chains, and thereby contributes to stiffness of the matrix. Affinity for water through oncotic, *i.e.*, osmotic swelling, effects also derives from these fixed ionic groups. It thus plays a role not only in the normal turgor of the tissue but also, likely too, in certain edematous states. Another important physiologic consequence is akin to the piezoelectric effect of crystals. When mechanical forces are applied to the tissues, the charged groups are displaced with respect to each other, and an electric potential is generated. They are in this way able to attract other charged molecules. This must be regarded as a fundamental mechanism whereby connective tissues take shape to meet mechanical requirements in normal development as well as disease.

The skeleton and teeth acquire sufficient stiffness for their function through a high content of mineral. It would be a mistake, however, to regard these structures simply as rigid materials. Like other connective tissues, bone is viscoelastic, *i.e.*, it deforms in non-linear manner when subjected to mechanical stress over a period of time (Fig. 1.1*B*). The viscous component, by flowing, absorbs energy and thereby cushions the shock of impacts which might otherwise cause some of it to break. In bone, the viscous component resides in the organic matrix. The disposition and attachment of the mineral crystallites to the organic matrix allow the bone to have the right mix of stiffness and of strength in shear. Compare, for example, ceramic materials which are composed of mineral crystallites. Although strong in compression, ceramics break readily in shear or tension because the crystallites are readily torn apart from each other. Hardness thus is by no means identical with strength. Hard material may be very brittle because of inhomogeneities. Mechanical stresses characteristically concentrate (stress = force/unit area) along minor discontinuities such as cement lines and so cause disruption of the part. It is thus not surprising that fractures occur frequently in bone diseases where mineral material is excessive, *e.g.*, osteopetrosis, as well as those where it is deficient (osteopenias).

CHEMISTRY OF CONNECTIVE TISSUE MATRIX

Although this term is employed in various senses, the matrix is considered here to be the entire extracellular portion of the connective tissues.

Collagen

Collagen is the most abundant protein in the body, constituting approximately 30 percent of the whole. Its functional importance is related to its strength and stability. Although collagen is synthesized and secreted from cells into the surrounding spaces in soluble form, the molecules become converted there into insoluble fibrils and fibers by two distinct mechanisms: there are electrostatic and therefore largely reversible aggregations, and more stable forms arising from intermolecular chemical cross-links. Undenatured collagens resist digestion by ordinary proteases and highly selective enzymatic mechanisms (collagenases) are required for their physiologic turnover.

Collagen is actually not one but a series of similar proteins characterized by their amino acid composition and arrangement in a triple helix. These proteins are synthesized largely by connective tissue cells (fibroblasts, chondroblasts, and osteoblasts) but also by other cell types. In neurofibromas, tumors of neural origin, large quantities of collagen are present. There is an enormous amount of information about the chemistry of collagen. For present purposes, we indicate only several features.

The elementary polypeptides of collagen exist in the form of elongated strands called α-chains. The amino acid content and primary structure of α-chains vary in different tissues, but, in all, every third amino acid is glycine. Two of the other amino acids, hydroxyproline and hydroxylysine, although constituting but a small proportion (14 and 1 percent, respectively) of the total are, with a single exception, unique in collagen and important in its molecular configuration. The α-chains are formed into three-stranded cables, united to each other by covalent, intramolecular cross-links.

Synthesis of collagen originates on ribosomes where a precursor, protocollagen, is formed. In protocollagen, there is no hydroxyproline or hydroxylysine but only corresponding proline and lysine. The conversion to hydroxyproline and hydroxylysine is a post-translational event, which requires, in addition to specific proline and lysine oxidases, ascorbic acid. The defect in collagen formation that characterizes scurvy is the consequence of the failure to oxidize the protocollagen attending avitaminosis C.

The monomer of the fully formed molecule, tropocollagen (mol. wt. = 300,000), measures 3000 Å in length and 15 Å in width. It is water soluble. Formation of microfilaments results initially from electrostatic aggregation. A progressive hierarchy of side-to-side aggregation comes about through covalent intermolecular cross-links. Fibrils are aggregates of filaments that are not separated from each other by ground substance in electron micrographs: their width varies from 0.01 to 0.1 μ. Fibers, by contrast, are larger (0.1-mm diameter) bundles of fibrils that are separated from each other by ground substance. Fibers occur in certain connective tissues and not in others. They are best seen in tendons and ligaments where they can be detected by the naked eye. Fibers in this sense do not occur in joint cartilage: in this tissue the widest collagen group is the fibril.

Collagen has a characteristic striated appearance under the electron microscope, a feature useful both in morphologic recognition of fiber and in the analysis of its molecular structure (Fig. 1.2). Major periods are approximately 640 Å in length and a series of smaller internal bands are distinguished within these. The length of these major periods is only approximately one-fourth that of the tropocollagen molecule indicated above. To account for this, the "quarter-stagger" arrangement of the monomeric collagen is generally accepted. According to this model of the fibril, the collagen molecules are aligned side-by-side with each other so that one overlaps with its neighbor by one-quarter of its length. A small gap (approximately 400 Å) separates the actual ends of the molecules from each other. This "hole" is significant because it may be a site for the nucleation of calcium salts in the early mineralization of collagenous tissues. Under laboratory conditions, tropocollagen molecules can be made to line up with each other in register rather than in quarter-stagger. The periods then are 2700 Å rather than 640 Å long; hence the name, long-spacing collagen. Long-spacing fibers, however, are encountered only infrequently in tissues.

Figure 1.2. Electronmicrograph of collagen in the radial layer of human articular cartilage. The fibrils are discrete and arranged randomly in the ground substance. Their diameters vary from 100 to 700 Å and the characteristic cross-banding is seen in the widest of them. (×30,000.) (Courtesy of Dr. Aubrey J. Hough.)

6 Musculoskeletal System

Stability of the side-to-side attachments is a molecular prerequisite for strength of the collagen and is achieved by the covalent intermolecular cross-links. The major if not exclusive cross-link involves lysine or hydroxylysine of adjacent molecules. The first step in cross-linking is a slow enzymatic deamination of the lysyl residues. Copper is a component of the amino-oxidase. The resultant aldehydic group is available for condensation of the two molecules. Defective cross-linking through failure of the deamination of the lysyl residue occurs in several pathologic states. Osteolathyrism is a disorder that results from ingestion of a component (β-aminopropionitrile) of a sweet pea (*Lathyrus odoratus*) meal. As a result the skin is highly friable, the skeleton deforms and the aorta is prone to rupture. Marfan's syndrome has a certain similarity to this as do copper deficiency states. Homocystinuria is an inherited disorder in infants. Excessive quantities of homocystine are formed and these react with the newly elaborated aldehydic groups, interrupting the condensations.

Collagen contains a small quantity of sugar (galactose or glucosylgalactose) linked glycosidically to hydroxylysyl residues. In reticulin, the delicate fiber that forms the sheath of skeletal muscle fibers as well as the framework of lymph nodes and spleen, hexoses constitute 4 to 10 percent of the weight instead of the 0.5 percent found in other collagens. These sugar moieties must be responsible for the characteristic staining of reticulin by special silver and periodic acid-Schiff reagents.

The molecular species of collagen in cartilage is different from that in bone and skin. Endochondral ossification, the process by which a preliminary cartilage is replaced by bone during maturation of the skeleton, cannot thus be based on a reutilization of the cartilage collagen. The latter must first be removed and then replaced by synthesis of a bone type of collagen. The turnover of collagen under all circumstances is low as measured by the levels of hydroxyproline-containing peptides in the body fluid and urine. The values are somewhat higher in growing children than adults. There are lysosomal collagenases in many vertebrate cells. At least two types of collagenase are present in polymorphonuclear leukocytes which are involved in the lysis of collagen during suppuration. The collagenase of *Clostridium perfringens*, the principal organism of gas gangrene, has a broader spectrum of collagen substrates than the others and contributes to the spread of the inflammation that characterizes the infection. When collagen is heated, even to relatively low temperatures, it undergoes profound molecular disorganization. Exposure to 68°C for even 1 minute results in a marked and irreversible shrinkage. Comparable changes occur at lower temperatures when the period of exposure is longer. More drastic heating results in denaturation of collagen into gelatin. Gelatin, unlike collagen, is readily destroyed by many proteases.

Because collagen, once laid down, remains in the tissues indefinitely, it would seem a natural target for molecular mechanisms of aging. From what has just been said, such mechanisms might include excessive cross-linking or thermo-oxidative transformations that would create major alterations in the physical properties of connective tissues. This type of consideration occupies much of the contemporary literature on experimental gerontology.

Ground Substance

The nomenclature of ground substance components changes from time to time and is apt to be confusing. The components are a heterogeneous group of macromolecules variously called connective tissue mucins, mucoproteins, proteinpolysaccharides, or proteoglycans. In these macromolecules, protein cores are linked covalently to special types of long-chain polysaccharides known as mucopolysaccharides or glycosaminoglycans. One or two sulfate groups are fixed to each repeating sugar unit in many mucopolysaccharides so that they constitute strongly electronegative polyelectrolytes. Although much is known about the chemistry of mucopolysaccharides, it must be emphasized that these molecules are analytical artifacts and do not exist as such in connective tissues. It is only in the physicochemical properties of the intact proteinpolysaccharides that the function of the ground substance can be understood.

Proteinpolysaccharides form viscous sols with water by virtue of their large size and electrostatic properties. Their hydrodynamic size in cartilage is further increased by a non-covalent interaction with another matrix protein, link protein. The precise nature of the interaction of ground substance with collagen is not known to everyone's satisfaction, but the result is that the ground substance is in effect cross-linked so that it behaves more like a gel than a sol. From this derives its principal functions:

1. As in chromatographic gel filtration, it constitutes a molecular sieve, admitting small molecules and excluding large ones from its domain.
2. By further analogy with chromatographic separation, it serves as an ion exchanger, retaining or excluding molecules according to their charges.
3. It has an oncotic pressure, causing turgor through imbibition of water and electrolytes.

These are the mechanisms by which ground substance contributes not only to the support function of connective tissues but also regulates the transport of water and large and charged molecules throughout the body.

Mucopolysaccharides are linearly polymerized. One moiety of each monomer is an acetylated hexosamine (glucosamine or galactosamine) and the other is a uronic (glucuronic or iduronic) acid. Keratan sulfate, in which galactose replaces the uronic acid, is nevertheless usually classified as a mucopolysaccharide. Sulfate, when present, is attached to the 4 or 6 carbon of the hexosamine. The location and degree of sulfation affect the above mentioned functions of the ground substance and also the turnover rate. A small peptide chain also is an intrinsic part of the mucopolysaccharide. Its importance resides in the nature of the attachment of the mucopolysaccharide to its protein core. This attachment is a point where proteases can split the core protein from the mucopolysaccharide in physiologic or disease states. In the case of keratan sulfate, at least, the junctional group of the mucopolysaccharides is a xylose-serine link. The microheterogeneity of the ground substance arises not only from the diversity of the types of polysaccharide and protein, but from their permutations and combinations within the chain. Sialic acids are derived from another amino sugar, neuraminic acid, which has 9 rather than 6 carbons. Although they occur in connective tissues in small quantities, they are more characteristic of epithelial mucins and glycoproteins. The principal mucopolysaccharides and the tissues in which they are most characteristically distributed is indicated in Table 1.1.

Mucopolysaccharides are synthesized by all sorts of connective tissue cells and are secreted promptly into the matrix. Unlike collagen, they turn over rapidly. The half-life of chondroitin sulfate in several connective tissues is of the order of 10 days. Lysosomal enzymes are involved in the degradation of these compounds. Inherited defects in these lysosomal enzymes result in the mucopolysaccharidoses, storage diseases in which the specific mucopolysaccharides accumulate within the cells (p. 75). Hyaluronidase, which can depolymerize hyaluronic acid and chondroitin sulfates, occurs in many tissues. Other polysaccharidases (β-glucuronidase, acetylhexosaminidase) occur in mammalian tissues; their role in the turnover of ground substance is not so clear. Alkyl sulfatases that can remove sulfate from mucopolysaccharides have not been identified in mammalian tissue. Mucopolysaccharides sulfated in the C6 position are more stable than those in C4. Presumably this is why keratan sulfate and chondroitin sulfate 6 increase disproportionally in many connective tissues of elderly persons. Lysozyme, although an enzyme in certain other systems, occurs in appreciable quantities in cartilage matrix. It has no apparent substrate in this tissue and its physiologic significance may

TABLE 1.1
Composition and Distribution of Mucopolysaccharides

Mucopolysaccharides	N-Acetyl-hexosamine	Uronic Acid	Hexose	Sulfate	Monomers per Chain (approximate)	Characteristic Tissue
Hyaluronate	Glucosamine	Glucuronic	—	—	2500	Synovial fluid, dermis, Wharton's jelly, vitreous humor
Chondroitin sulfate 4	Galactosamine	Glucuronic	—	C4	60	Cartilage, dermis, aorta, nucleus pulposus
Chondroitin sulfate 6	Galactosamine	Glucuronic	—	C6	60	Cartilage, dermis, nucleus pulposus
Dermatan sulfate	Galactosamine	Iduronic, some glucuronic	—	C4, some C6	30	Dermis, fibrocartilage
Keratan sulfate	Glucosamine	—	Galactose	C6	10–20	Cornea, cartilage
Heparin	Glucosamine	Glucuronic	—	C6, NH*	10–15	Mast cells

* In heparin, one of the sulfate groups replaces acetyl on the amino sugar; a third sulfate is present on half the uronic moieties.

be related to its high isoelectric point rather than an enzymic function.

The synthesis of ground substance components is profoundly influenced by hormones. Sexual dimorphism of the skeleton, acromegaly, endocrine dwarfism, and myxedema attest to this. Because of their complexity, further discussion is deferred until the specific disease entities are considered.

MICROSCOPIC APPEARANCE OF CONNECTIVE TISSUE MATRICES

The histologic character of the matrix varies not only with the quantity but also the organization of its contents. In some, the fibrous element is self-evident grossly and microscopically. In others, such as the corneal stroma, the matrix appears completely transparent despite a considerable collagen content. The explanation for this rests on the pattern of the collagen; in the case of the cornea, the fibrils are arranged in a perfect lattice. Hyaline cartilages are so-called because they appear glassy, *i.e.*, they are homogeneous and relatively translucent. In hyaline cartilages, collagen fibrils are copious but are thin and separated from each other by ground substance. They differ thereby from fibrocartilages in which the collagen is organized into bundles. Several histochemical and other microscopic techniques are used everyday in examining connective tissues.

Collagen, being fibrous, necessarily has a strong directional orientation, *i.e.*, anisotropy. This can be recognized in microscopic sections by its birefringence when examined with polarized light. Insertion of two crossed polaroid discs into the light path of a microscope is an inexpensive and easy way of accomplishing this. It should be done routinely in the histologic examination of connective tissues and in the clinical examination of synovial fluids for crystals.

Special stains commonly employed for collagen include the Masson trichrome and Van Gieson methods. With the Masson stain, collagen appears blue. In bone, unlike other connective tissue, for some reason, the matrix appears red. In Van Gieson preparations, mature collagen is red, while other materials stain yellow. In Wilder's and similar methods for staining reticulin, silver salts are reduced by the sugars under precisely controlled conditions; grains of silver are deposited on the fibers thus making them appear black. The periodic acid-Schiff stain involves the oxidation of vicinyl hydroxyl groups to aldehydes which in turn combine with the Schiff reagent. It thus stains glycosyl groups in reticulin, glycoproteins, and other polysaccharides but is little reactive with the isomers of chondroitin sulfate. Elastin is easily demonstrated by orcein, and a number of other stains are available.

Their polyanionic character permits several sorts of staining of ground substance components but little that is specific for one rather than another. The sulfated mucopolysaccharides are metachromatic, that is, they are stained off-color by cationic dyes such as methylene blue or toluidine blue; for example, the matrix of bone is stained blue, but the metachromatic cartilage appears violet under the same conditions. Hyaluronate, lacking sulfate, is not metachromatic. Another group of stains is based on the differential affinity of the acid mucopolysaccharides for quarternary ammonium dyes such as alcian blue under critical salt concentrations. Here again the analogy between the staining and the ion exchange chromatographic properties of the mucopolysaccharides seems appropriate. Safranin O also is a useful stain for chondroitin sulfate.

In mineralized connective tissues—bone, calcified cartilage, and dentin—satisfactory physical properties of the matrix require a preferred arrangement and cohesion between the crystallites and the collagen. Spacing between the crystallites in bone is necessary for a degree of flexibility. The elastic modulus of bone is intermediate between that of collagen and the mineral. Most of the bone salt is hydroxyapatite, $Ca_{10}(PO_4)_6(OH)_2$. The crystallites are needle-like, approximately 400 Å long, 30 Å wide, and thus much shorter than the collagen molecule. In bone matrix (the unmineralized component is called osteoid), some of the crystals are located within and are oriented in the axis of the collagen fibrils. Perhaps they lie in the "holes" between the ends of the linearly aggregated tropocollagen molecules. Some argue, however, that at least initially the bulk of the mineral lies external to the collagen.

In mature mammalian bone, collagen fiber bundles characteristically are arranged in alternating directions but the main axis is longitudinal. Fractures therefore occur more readily when the bone is torn across than with the fibrous grain. The lamellar pattern of the collagen is best seen with polarized light (Fig. 1.3). In immature or pathologic bone, the lamellar pattern is less defined; collagen bundles weave irregularly across each other in many directions, hence the name "woven bone" (Fig. 1.4). The water content of the osteoid decreases as mineral is deposited, necessarily embarrassing diffusion of water and small solutes between the osteocytes and the blood stream. To overcome this, bone matrix is traversed by a series of channels, the canaliculi, that communicate with the lacunae in which the osteocytes reside (Fig. 1.5). The canaliculi and lacunae are lined by a ground substance-like layer, sometimes called Newman's sheath.

In planning biopsies for histologic evaluation of mineralization, it is important that the specimen *not* be decalcified in the usual fashion prior to sectioning. Unfortunately this precaution often goes unheeded, and

Figure 1.3. Lamellar bone viewed with polarized light. The collagen-rich bundles are arranged in concentric rings approximately 7 μ thick about the Haversian canal. Straight rather than circular bundles (*arrow*) are interstitial lamellae. ($\times 350$.)

the diagnosis of osteomalacia is simply not possible for this reason. The Von Kossa stain ordinarily used to demonstrate calcium actually reacts with the phosphate or carbonate in the mineral deposits. Alizarin red, which stains the calcium proper, is often less satisfactory for technical reasons.

WATER AND SALT METABOLISM

In all connective tissues except bone, water is the largest component, most tissues containing upward of 70 percent by weight. Approximately one-third of all the body water is located in connective tissues and their function as a reservoir for homeostasis of other fluid components has received much attention. Water exists in three physical states of freedom: fluid, solid, and gel-like. The fluid type, water of simple solution, flows freely, as in plasma or lymph. A small component of connective tissue matrix (2.5 percent of all the body water) is of this type and designated transcellular water. It corresponds to lymph that has not yet found its way to lymphatic vessels. "Solid" water is the non-exchangeable water of hydration of minerals such as hydroxyapatite. The gel-like state accounts for the bulk of the water in connective tissue ma-

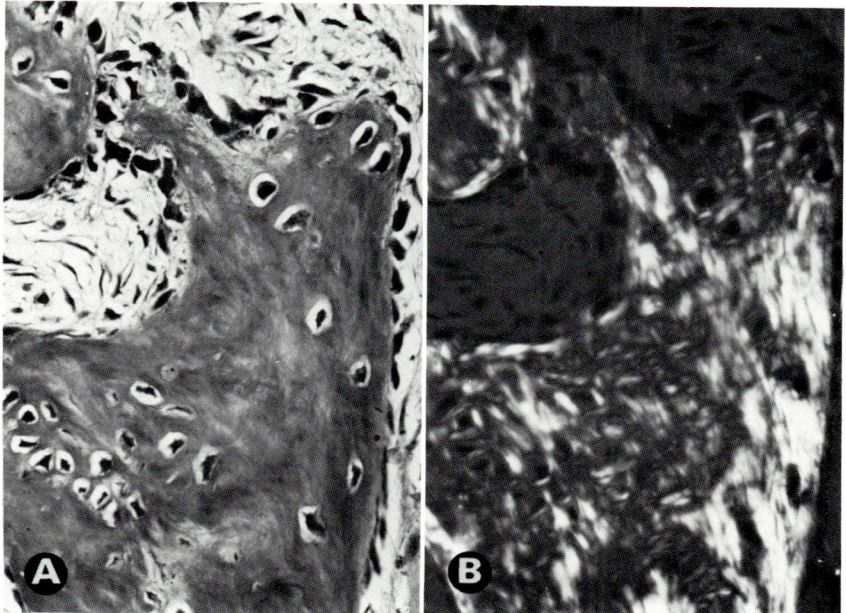

Figure 1.4. Membranous bone formation. *A.* Osteoblasts are present at the margins of the trabecula and also are present within the osteoid. *B.* When viewed with polarized light, the collagen trajectories have a woven appearance, crossing each other irregularly at right angles. Contrast with lamellar bone in Figure 1.3. (Hematoxylin and eosin, ×350.)

trices. Although some of the matrix water is held by the collagen, most is in the ground substance. As in other gels, the individual molecules of water and salt in the interstices of the macromolecules are readily exchangeable by diffusion with adjacent free fluid. Nevertheless, there is a constant net retention depending on the hydrostatic pressure and ionic environment. The fluid may be removed under artificial conditions such as evaporation, but in life this is accomplished by hydrostatic pressure (squeezing or suction) or alteration of the gel through changes in the electrolytes, pH, or cross-linking of the matrix molecules. An excess of cations, particularly polyvalent ones such as Ca^{++}, reduces the oncotic pressure and water content while the opposite effect results from a reduction in the cations.

Connective tissue matrices contain the bulk of all the body sodium, only 11 percent being in the plasma. Approximately half of the sodium is in the bone. Here much of it does not form an intrinsic part of the mineral crystallite but is held in a thin layer of water polarized about the ions at the surface of the crystallite, the hydration shell. Approximately 100 acres of crystallite surface—a small fraction of the total—are exposed to

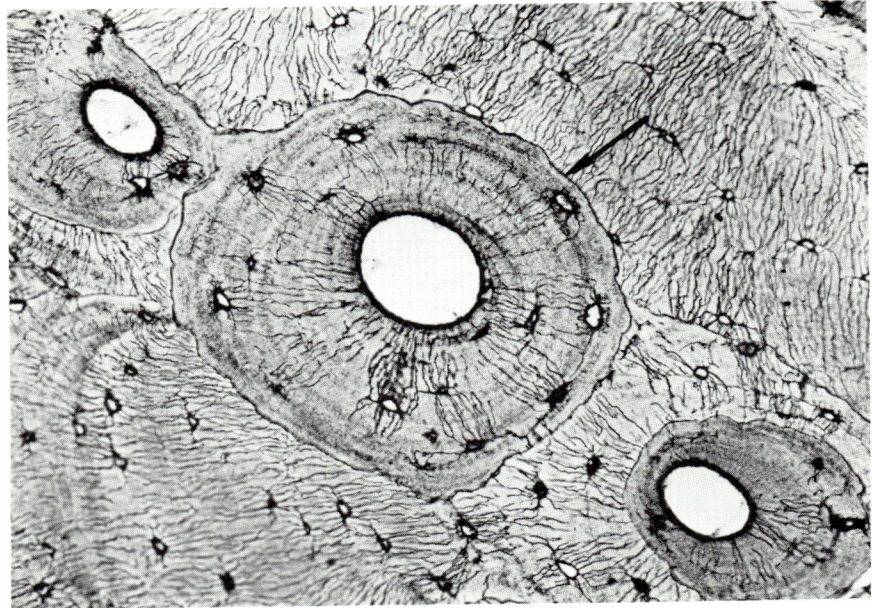

Figure 1.5. Nutritional pathways through bone matrix. The delicate "filaments" extending from the lacunae perpendicular to the lamellae are actually the canaliculi. The black stained material lining the lacunae is "Newman's sheath." The thin border at the periphery of the osteone (*arrow*) is the cement line. (Bodian, ×250.)

the circulating body fluids and available to the exchangeable pool. In experimental studies on rats, approximately one-third of the bone sodium is readily exchanged, the remainder sequestered more remotely in the matrix. Although some studies in man indicate that the exchangeable fraction is mobilized in acidosis, Pellegrino and Biltz found no reduction in bone sodium concentration in persons with longstanding uremia and presumably hyponatremia. It is difficult to reconcile these conflicting findings but we suggest that the large size of the exchangeable pool in the experimental animals resulted from their youthfulness and therefore more hydrated skeleton.

Some other cations do, however, enter and become immobilized in the crystal lattice. This is the basis for the use of short-lived bone-seeking isotopes (*e.g.*, ^{47}Ca and ^{87}Sr) in radiologic studies of the skeleton. ^{99m}Tc (technetium), now commonly employed in clinical bone scanning, is localized in reticuloendothelial cells in soft tissue defects rather than the mineral. Inclusion of exceptional cations in the mineral also gives rise to important diseases. By virtue of their long residence, radium and ^{90}Sr (strontium) cause malignant bone tumors: osteosarcomas. The former was at one time a significant occupational hazard; the latter results from nuclear fission fall-out. In lead poisoning, deposits of lead-containing crystallites appear as radiopaque zones in x-ray films of the bones, particularly at the growing epiphyses (Fig. 1.6). Among the anions that may enter the mineral lattice is fluoride. This is the basis for the fluoridation of water in preventing dental caries. Enormous excesses of fluoride ingestion in man also cause osteofluorosis which is characterized by excessive bone formation. Its therapeutic implications for skeletal disease are debated. The mechanism by which fluoride deposition may stimulate new bone formation is unknown. This problem is not unique to osteofluorosis; the signal for osteogenesis or bony metaplasia following other forms of pathologic mineralization, e.g., dystrophic calcification, also is enigmatic.

Only 1 percent of the total body calcium, a physiologically important 1 percent, exists outside of bone. The critical role of this ion in neuromuscular and other membrane functions requires that its concentration in body fluids remains constant. The partition with the calcium in the bones is thus central to the metabolism of skeletal tissue.

THE CELLS OF CONNECTIVE TISSUES

Connective tissue cells arise from mesenchyme in fetal life, and from fibroblasts and granulation tissue in postnatal life. In such circum-

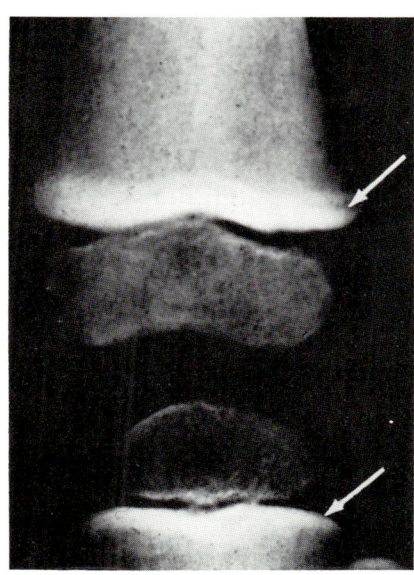

Figure 1.6. Lead poisoning, knee joint of 2-year-old child. The marked radiopacity (arrows) adjacent to the growth plates reflects the energy-absorbing capacity of Pb in the crystallites of the matrix.

stances, the parent cells have a fairly uniform and undifferentiated appearance. The fact that diverse cell types are derived from the apparently homogeneous primordium underlies the concept of the pluripotentiality of connective tissue. Because each matrix contains collagen and ground substance, it was long believed that development of mature connective tissue represents a modulation rather than a differentiation from a common stem cell; *i.e.*, the various tissues differ in the quantity rather than the kinds of their constituents. As a corollary, one cell type might convert into another when local conditions change. This concept is no longer tenable because it is now known that there are different molecular species of collagen. The epigenetic programming of chondrocytes accordingly is different from that of osteocytes and skin fibrocytes. This brings into focus the question whether mesenchymal cells or fibroblasts are individually pluripotential or, by analogy with lymphocytes, are committed to specific pathways. By this account, only certain cells would be slected to proliferate by their microenvironment. In either case, cellular proliferation in bones and joints characteristically leads to a variable but intimate admixture of hyaline and fibrocartilage, bone, and fibrous and synovial tissues. This histogenetic theme recurs not only in pathologic but therapeutic considerations of the musculoskeletal system. The nature of the stimulus for one path of evolution rather than another is not really understood. Anoxia may favor formation of cartilage which has a predominantly glycolytic metabolism and low requirements for oxygen, while bone formation is more likely where a greater blood supply is available. Mechanical factors undoubtedly are important. These factors must be recognized in restoration of musculoskeletal tissues following injury. Even ideal surgical repairs fail unless appropriate alignment and exercises are maintained during the period of cell growth. If they are not, a disorderly mixture of scar and callus is formed rather than differentiated tissues having the required degrees of mobility and strength. The piezoelectric mechanism for converting the mechanical into a biologic signal suggested above may operate here.

Mesenchymal cells are stellate and have long, slender cytoplasmic processes. *Fibroblasts* are rather similar although with maturation they become elongated and bipolar, the processes located at the ends of the cells. The configuration of other connective tissue cells departs from this. *Chondrocytes* are more rounded; short processes extend from the border of these cells within their lacunae. Mitochondria are less prominent than in other connective tissue cells. During calcification of cartilage, minute membrane-bound vesicles bud off from the periphery of the cells into the adjacent matrix. The vesicles are not lysosomal. They contain calcium crystals and this suggests active transport in the mechanism of mineralization. *Osteoblasts*, proliferating bone cells, are cuboidal and have a basophilic cytoplasm (Fig. 1.4). This appearance results from a large quantity of ribosomal RNA, some attached to the protein-exporting rough endoplasmic reticulum and some not. Alkaline phosphatase also is prominent in the cytoplasm. It is responsible for the high serum levels of alkaline phosphatase in growing children and in pathologic circumstances of new bone formation. It is distinguished from hepatic alkaline phosphatase by heat lability. Other morphologic evidence for a role of osteoblasts in calcification is scanty. These features disappear from the mature *osteocytes*. The latter may be regarded as resting cells. Cytoplasmic processes are better developed in osteocytes than in chondrocytes. They extend into the canaliculi but there is no evidence that they establish cell-to-cell contact thereby *Osteoclasts* are multinucleated, motile cells now known to destroy mineralized bone (Fig. 1.7). They do not act on osteoid that has not been calcified. These cells have from 15 to 20 uniform nuclei and are located in chewed-out looking pits in the margins of the bone called Howship's lacunae. Their precise origin is debated. Although they may develop from osteocytes, the fine structure

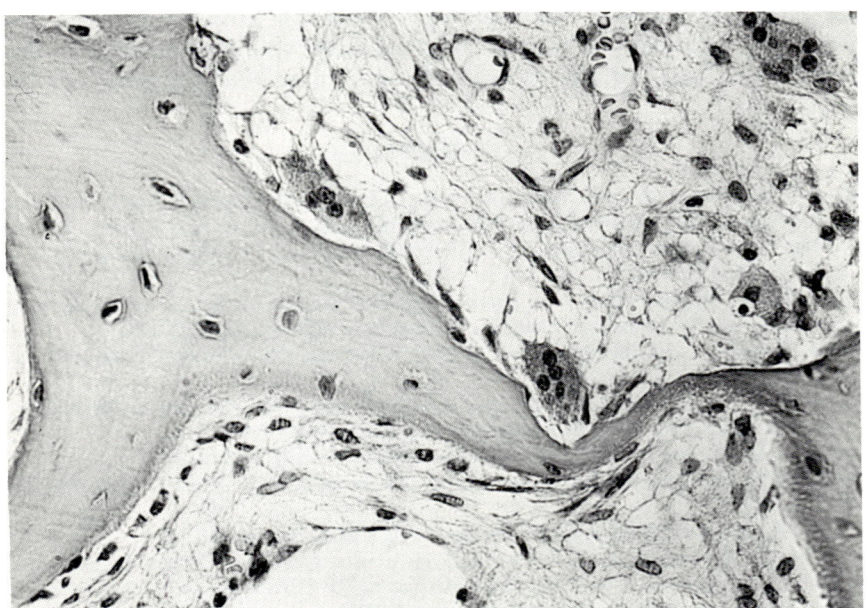

Figure 1.7. Osteoclastic resorption of bone in secondary hyperparathyroidism. The thickness of the trabecula has been reduced to a slender thread in the vicinity of the multinucleated cells. The adjacent marrow has become fibrous, as is characteristic of this condition; hence the name osteitis fibrosa. (Hematoxylin and eosin, ×350.)

2

BONE AS TISSUE AND ORGAN

A distinction should be made between bone as tissue and organ. Bone as organ contains not only osseous but marrow tissue. These compartments vary in disease while the volume remains constant. Values presented for the composition, physical properties, and thickness necessarily vary according to the basis for measurement employed. The following figures for the composition of bone are based on bone as tissue. Water constitutes approximately 60 percent of young osteoid tissue but only 8 percent of mature mineralized bone. Of the dry weight, 76 to 77 percent is mineral and the remainder is organic. Ninety percent of the organic material is collagen; 2 percent is sialic acid-containing and other glycoprotein. Only 1 percent is sulfated proteoglycan. The remainder comprises miscellaneous cellular and blood constituents.

GROWTH AND DEVELOPMENT

Bone always forms in pre-existing connective tissue. When the latter is cartilaginous, the process is called endochondral ossification; when the pre-existing connective tissue is non-cartilaginous, ossification is membranous. Membranous ossification occurs principally in the cranium while endochondral ossification is characteristic of most of the skeleton. The latter statement is only partly true because while longitudinal growth requires a preliminary proliferation of cartilage, transverse growth of the shaft takes place in the periosteum. This is significant clinically because the latter phenomenon allows a slow but distinct increase in the diameter of the bones in adult life even after the body has achieved its full height.

The general configuration of the cartilaginous skeleton is determined genetically, but even in fetal life bony development may be modified by mechanical forces. The ends of long bones are generally broader than the shafts. This allows mechanical loads to be distributed over a broader surface and thereby reduce the stresses on joints. Furthermore, by locating tendons farther from the pivotal center of rotation of the joint, the mechanical advantage of the muscles is increased. In the shaft (diaphysis) the bone is arranged in a thick and compact cortex. Little bone is present in the adjacent marrow. Toward the ends, bone is disposed in a more finely divided spongy or cancellous system of struts, the trabeculae. The conversion of the original cartilage model into this more elaborate structure requires a complex process of resorption and laying down of new bone (remodeling). In endochondral ossification, the cartilaginous template is resorbed locally and replaced by a newly formed bone matrix. Two morphologic events precede this: calcification of the cartilage matrix and ingrowth of blood vessels. Chondroclasts, analogous to osteoclasts, participate in the removal of the calcified matrix. A layer of osteoid is deposited on the original cartilage model by osteoblasts. At first the bone matrix is woven; later it is replaced by a lamellar type.

In tubular bones, ossification begins toward the center of the shaft, the ends (epiphyses) remaining cartilaginous so that growth can occur. Secondary centers of ossification appear in the epiphysis at a later date and enlarge progressively. When the primary and secondary centers of ossification unite with each other, longitudinal growth of the skeleton is complete. The time at which these events occur varies from bone to bone. There are local as well as genetic determinants of ossification. Local injuries or inflammation of the growth cartilage may cause vascularization and thereby premature ossification with restricted growth of the part. In addition, however, the rates at which these events occur throughout the body are profoundly influenced by hormonal and other regulatory mechanisms. The patterns of ossification and epiphyseal union as seen in x-ray films for this reason are widely employed as a gauge of "bone age" in pediatric evaluation of growth and development.

Cartilaginous growth of the tubular bones takes place in the epiphyseal plate, a relatively narrow (2 mm or so thick) zone at the base of the epiphysis. The junction of the epiphysis with the shaft of the bone constitutes the metaphysis. In the epiphyseal growth plate, chondrocytes are arranged in longitudinal columns. Cell division results in linear elongation of the bone but there is a lateral expansion of the epiphysis, too. A thin, discontinuous rim of periosteal bone extends onto the margin of the epiphyseal plate. This is the periosteal ring, noteworthy because it constrains lateral growth at the plate and constitutes a locus minoris resistentiae in infectious joint disease. If the increased diameter of the growing epiphysis persisted, the configuration of the bone would become distorted. Distortion is prevented by osteoclastic cut-back of the excessive bone from the periosteal aspect of the metaphysis (Fig. 2.1). Numerous skeletal anomalies arise from specific defects in this remodeling process. Failure of the metaphyseal cut-back mechanism, *e.g.*, in metaphyseal dysplasia, results in bulbous enlargement of the metaph-

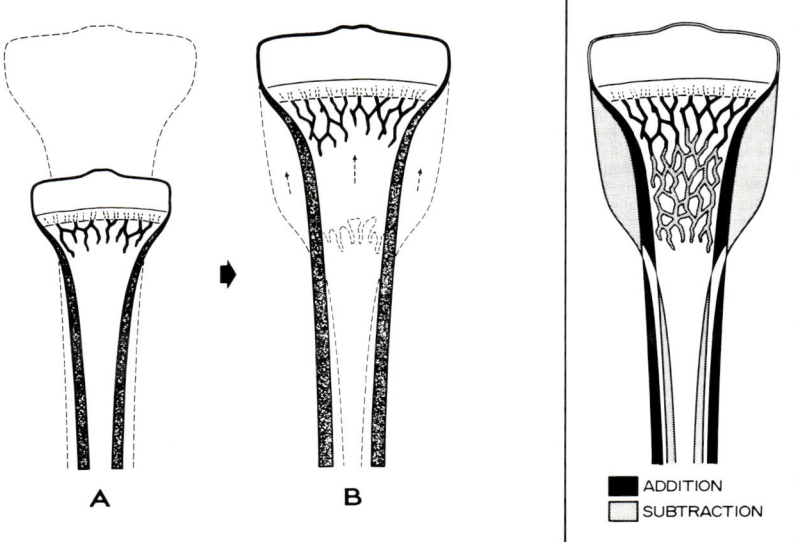

Figure 2.1. For a bone to have a normal shape as it grows, remodeling is necessary. Bone must be removed as well as added, and the sum of the two processes results in the calcium balance.

ysis. Resorption of the calcified cartilage leads to the formation of bone and of marrow spaces. The interface between the marrow connective tissue and the bone constitutes the endosteum. The latter is not a discrete cellular layer. Endosteal remodeling is the basis for cancellous bone formation. As bones grow in width, they enlarge on the external aspect by membranous ossification of the periosteum. Two layers of periosteum are distinguished: an inner cambium layer from which new bone is generated and an external fibrous layer which serves as a mechanical envelope.

BONE BIODYNAMICS

Bone, in its mature configuration, is not static but undergoes throughout life a continuous turnover through two separate modalities, physicochemical and cellular. There is an exchangeable pool, a fraction of bone salt that is in dynamic physical equilibrium with body fluids. Exchanges between the fluids and the mineral occur rapidly at the interface and necessarily therefore along canaliculi and the margins of the lacunae. The other mechanism is a slower structural turnover of a stable compartment involving osteoclastic removal of the organic as well as the inorganic matrix. The resorption is coupled with a proliferation of blood vessels which invade the matrix. In turn, it is followed by a concentric laying down of new bone about the blood vessels, thereby forming secondary osteones (Haversian systems). The alternating resorption and accretion of bone represent a remodeling of the adult skeleton. It is a mechanism allowing bone to adapt to changing mechanical requirements and it also provides a blood supply for bone cells as the cortex widens. Although osteones are often described as the functional and structural units of bone, they are not a feature of all bones or inborn within them. Haversian systems occur in cortical rather than trabecular bone. They develop at various locations and appear mostly during adult life. Cortical bone is converted as a result from a laminar into a porous structure (Fig. 2.2.). The process takes place principally at the endosteal surface and proceeds toward the periphery of the cortex. The laminar bone persisting between the secondary osteones constitutes the interstitial lamellae. In addition, the participation of osteocytes in the resorption of bone under certain circumstances is increasingly recognized. This results in enlarge-

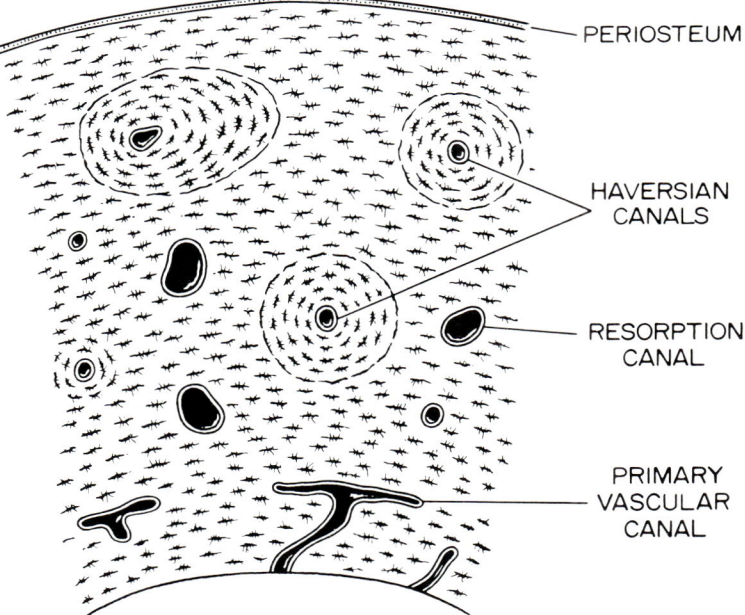

Figure 2.2. Remodeling of cortical bone as seen in transverse section. The original bone is laminated parallel to the periosteum and its blood vessels course in the same plane. The first step in remodeling is an osteoclastic tunneling and ingrowth of blood vessels in the long axis of the cortex (resorption canals). New bone is then laid down in concentric lamellae about the latter vessels, thereby creating Haversian canals.

ment of lacunae (osteocytic osteolysis).

The course of the remodeling is traced in metabolic bone disease by two biopsy techniques. In microradiographs of undecalcified bone (Fig. 2.3), resorption of mineral is made visible by loss of radiopacity. The moment of deposition of calcium in newly mineralizing matrix is identified by the affinity of the ossifying front for tetracycline or its analogues. Tetracycline persists indefinitely in these spots and is demonstrated there by its fluorescence. From the time interval between pulses of these fluorochromes and the distance between them in the sections, the rate of bone formation can be computed. By these and other means, the rate of osteoclastic resorption is estimated roughly as 100 μ a day, while new bone fills in the resorption cavity only 1 μ in the same period of time. The amount of bone turned over structurally is about 2 or 3 percent per year in adults. The net *balance* between addition and subtraction is an increase during the first 5 decades of life; thereafter the amount of bone is reduced at a rate of 5 to 10 percent per decade. The rate of loss or at least the consequences of the loss is greater in women than men. These structural changes have chemical counterparts in metabolic balance.

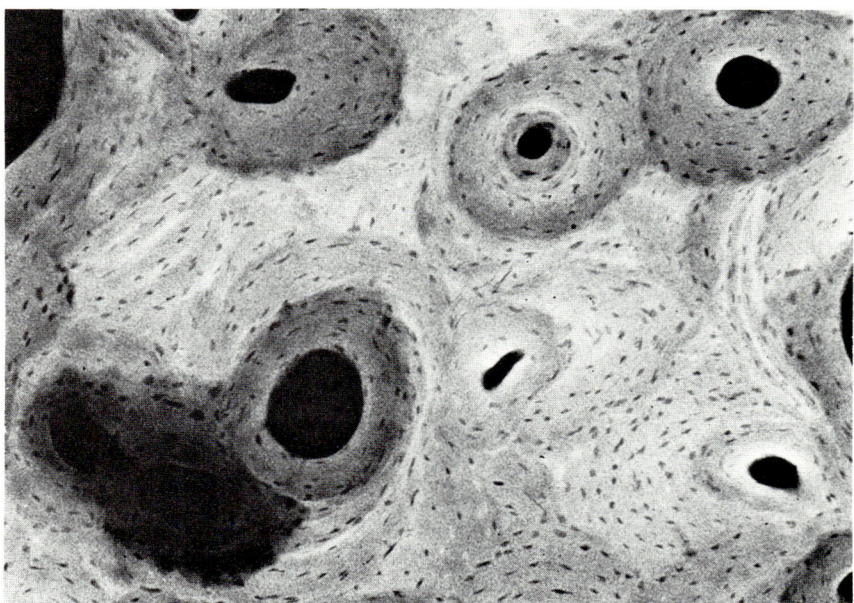

Figure 2.3. Microradiograph showing uneven radiodensity (*light areas*) of the matrix. The osteones having the largest dark space about the Haversian canals and least mineralization elsewhere are the most recently formed and metabolically active. Much of the remainder is removed from the metabolically active pool. ($\times 88$.)

Many factors enter into the remodeling and development of Haversian systems in cortical bone. Aside from genetic and other regulatory mechanisms, proper recognition of mechanical stimuli must be made. This is often expressed as Wolff's law which, although formulated in several ways, states that all changes in the mechanical function of a bone are attended by definite alterations in its structure. It has already been pointed out that piezoelectricity may be the transducer which converts the mechanical stimulus into chemical signals altering the configuration of the tissue.

VASCULAR AND NERVE SUPPLY

The circulation of blood occupies a central position in the physiology of bone. In laboratory animals approximately one-fourth of the cardiac output goes to the skeleton exclusive of its marrow, and the cortical blood flow rate is 18 ml per 100 g bone per minute. Extrapolated to the standard 70 kg man, this would amount to 1400 ml per minute, and thus be comparable to the renal circulation in the ionic homeostasis of the body. The figure obtained in anesthetized man, however, is much lower—only 5 percent of the cardiac output.

The routing of blood is intimately related to patterns of growth, remodeling, and disease (Fig. 2.4). Most of the blood supply to the cortex of the shaft comes by way of medullary vessels; periosteal vessels contribute little to this. Medullary vessels in tubular bones are derived from only one or two nutrient arteries which extend within the shaft toward the ends of the bones. In childhood, before the epiphyses have united, the terminal branches of the nutrient arteries at the epiphyseal plate loop back toward the venous circulation and do not anastomose with other vessels. This arrangement of the metaphyseal vessels is related to the regulation of ossification of the epiphyseal plate. Unfortunately, it also predisposes to impaction of infectious agents, and osteomyelitis most frequently starts in this area. Cortical bone is supplied exclusively with capillaries, a design suited to ensuring circulatory exchange for all the osteocytes. Transverse passages in the cortex for blood vessels are commonly called Volkmann canals. Arteriolar pressure is higher in the medulla than in extraskeletal soft tissues and blood is driven thereby from endosteum toward the periosteum. The cortical capillaries return to the marrow cavity where they enter into a large sinusoidal system and thence to nutrient veins.

By contrast with the shaft, the epiphyseal cortex is perforated by the numerous arteries which anastomose extensively with vessels of the joints and the adjacent periosteum. Communication with nutrient vessels is established in association with ossific union of the epiphyses. Little is known of the control of the circulation in bone. Poorly

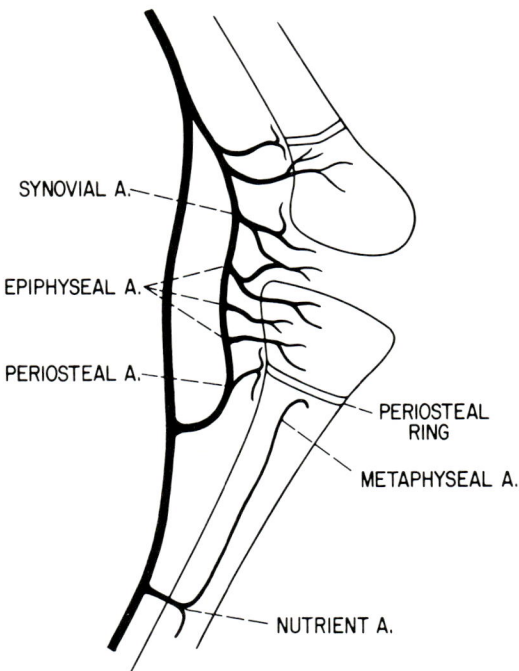

Figure 2.4. Schema of arterial circulation of bone and joint. A single nutrient artery supplies the diaphysis. Its metaphyseal branches, prior to closure of the epiphyseal plate, are end-arteries and do not communicate with those of the epiphysis. The epiphysis receives numerous arteries which anastomose with those of the apposed epiphysis, the joint and adjacent periosteal ring.

myelinated nerve fibers are present in Haversian canals. They may have vasomotor functions. There is evidence that a lowered pH in skeletal tissue associated with slowing of the circulation favors formation of new bone; while an alkaline drift resulting from accelerated blood flow leads to resorption. Most pain sensibility is located in the periosteum. Patients commonly complain of discomfort during aspiration biopsy of bone marrow, suggesting that medullary vessels also are innervated. Lymphatic channels are present in periosteum but not within the cortex or medulla of bone. Infection or tumor metastasis in these structures must therefore be blood-borne.

MECHANISM OF MINERALIZATION

Mineral constitutes approximately three-fourths of the dry weight and one-fourth of the volume of adult bone matrix. Deposition of the bone salt is a physicochemical phenomenon but its ultimate mechanism is unknown. Both humoral and local factors are involved. In disease states, the serum Ca and P concentrations generally vary in opposite directions. The normal Ca concentration is 10 (range 9.2 to 10.5) mg per 100 ml. Approximately half of this is ionized, but the figure varies when serum proteins which bind Ca are disturbed. Serum phosphate exists in three ionic forms (ortho, meta, and para), the proportion among them depending on the pH. The values are therefore expressed as inorganic phosphorus. Normally this is 3.8 mg per 100 ml in adults but the figure is far more variable than that of the ionized calcium. Bone mineralization becomes defective when the Ca concentration or particularly when the Ca $\times$ P product is low. Use of the Ca $\times$ P product no longer enjoys much clinical favor because it indicates only that certain minimal concentrations in tissue fluids are required for mineral deposition. Nevertheless, the preceding statement remains valid empirically. Furthermore, all major types of rickets and osteomalacia (excluding those associated with renal failure) are reversed by elevating the product, no matter what the means.

The local factors are both cellular and physicochemical. One possible mechanism in the precipitation of the salts is a local elevation of the ion concentrations so that their solubility product is exceeded. The role of the alkaline phosphatase in this regard is unknown; it is not simply to supply more phosphate ions by hydrolyzing phosphate esters. By comparison with ordinary solutions, extracellular fluid is already supersaturated with respect to calcium and phosphate ions. The concentration of these ions is further greatly increased in the hydration shell about the depositing crystallites. There are other difficulties with the simple mass action explanation for mineralization. The bulk of the mineral, for one thing, is not amorphous calcium phosphate but hydroxyapatite. There is no hard evidence that osteoblasts or osteocytes contribute directly to the mineralizing process, and vesicles comparable to those about chondrocytes in calcifying cartilage are less well established for bone cells. The chemical properties of the inorganic matrix are therefore involved. Precipitation of crystallites from the supersaturated state might result from "seeding" with another appropriately spaced molecule. The fact that crystallites in bone lie within and in register with the cross bands of the collagen fibers suggests to some investigators that polar amino acids of the bone collagen chains seed the hydroxyapatite from the extracellular fluid. To others, the previously described initial location of the crystallites between rather than in the fibers indicates that ground substance proteoglycans or phospholipids are involved. The role of pyrophosphatase, adenosine triphosphate, and citrate, all of which occur in bone, in the mineralization process is cloudy. In chondrocalcinosis, pyrophosphate rather than hydroxyapatite is deposited in joint tissue.

REGULATION OF MINERALIZATION

A complex regulatory mechanism involving the gastrointestinal tract, liver, kidney, and endocrine system maintains the serum Ca^{++} level (Fig. 2.5). Foremost among these is the parathyroid hormone (PTH). The rate of secretion of PTH is controlled by the serum Ca^{++} concentration through a negative feedback mechanism. A fall in the Ca^{++} level directly stimulates the parathyroid glands to synthesize and release PTH. The response is prompt and readily measured in the serum by radioimmunoassay. Reciprocally, elevation of serum Ca^{++} reduces the PTH concentration. Little preformed hormone is stored in the parathyroid gland and the half-life of PTH is about 10 minutes. Phosphate has no known direct effect on PTH secretion. PTH exerts two separate actions in bone metabolism: it causes osteoclastic resorption and it increases phosphate excretion by the kidney. Both actions are mediated at the cellular level by cyclic AMP (3′:5′-adenosine monophosphate). The effect of PTH on bone is to release the organic as well as the inorganic matrix; as a result PTH increases Ca and hydroxyproline levels in serum and urine.

Calcitonin (CT) is a small peptide hormone secreted by the parafollicular C cells of the thyroid gland. It inhibits osteoclastic resorption of bone and lowers the serum Ca concentration. Its net effect thus counteracts PTH but CT is not simply a parathyroid antagonist.

Vitamin D was originally recognized for its antirachitic action but now is known to function in calcium homeostasis over a broad spectrum of diseases. It is in one sense a vitamin or provitamin and, in another, a hormone or prohormone. Vitamin D can be synthesized from ingested plant ergosterol by ultraviolet irradiation of the skin, but it achieves its full physiologic activity by being itself converted into 1,25-dihydroxycholecalciferol (1,25-$(OH)_2D_3$). Vitamin D, in its natural form (D_3, cholecalciferol), has one hydroxyl group added at the 25 carbon position in the liver. The final step is the addition, in the kidney, of a second OH at the 1 carbon. The role of PTH in regulating the synthesis of 1,25-$(OH)_2D_3$ in the kidney is unsettled. 1,25-$(OH)_2D_3$ circulates through the blood as a hormone and regulates the transport of Ca^{++} through cells. Its role in preventing rickets is related to formation of a transport protein in the intestinal epithelium which absorbs Ca^{++} from the lumen and allows it to enter the blood stream. There seems to be an intimate interaction between PTH and 1,25-$(OH)_2D_3$. The bone-resorptive action of PTH requires that the latter be present. Like PTH, 1,25-$(OH)_2D_3$ acts on a number of cells including bone cells. Curiously in bone it acts like PTH to resorb tissue; its net bone-mineralizing effect in rickets results from the countervailing increase of serum Ca from the

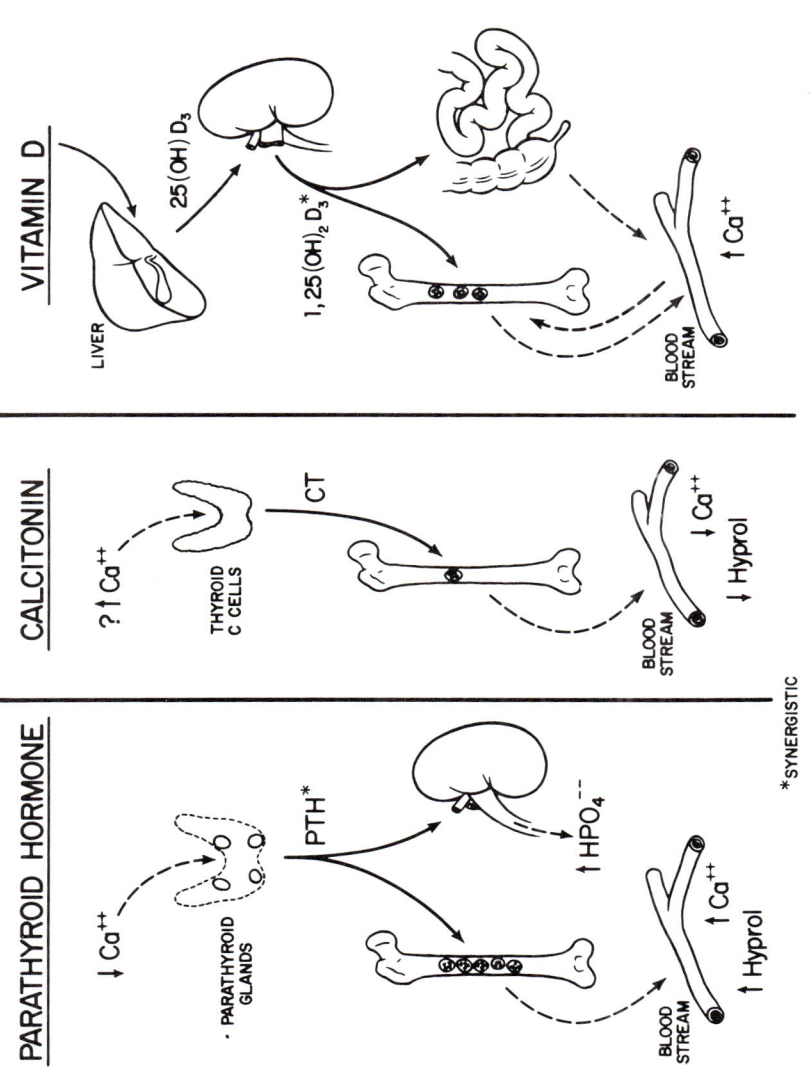

Figure 2.5. Three-part hormonal regulation of mineralization. The *broken lines* indicate the movement of Ca^{++}. The "smiling faces" in the bone represent osteoclasts.

intestine. Rapid advances are being made in the recognition of calcium disturbances resulting from faulty metabolism of vitamin D in liver and, particularly, kidney diseases.

Severe dietary deficiencies of calcium not only cause disorders of skeletal growth in the young but also lead to atrophy of bone in the adult (osteoporosis), particularly when special demands are made on the skeleton, e.g., during lactation. These deficiencies are exaggerated when the diet contains a large amount of phosphate or calcium-binding compounds such as phytate. The generally recommended intake of calcium is approximately 1 g per day. This figure is based on data from growing experimental animals and is probably excessive. The skeleton is able to retain calcium in the face of even prolonged dietary restrictions and also can tolerate wide variations in Ca:P ratio of the food, provided that vitamin D is available. Another factor that enters into the calcium balance is the endogenous excretion of calcium into the gut. There is a two-way traffic for calcium between the blood stream and intestinal lumen. Under ordinary circumstances, some 11 liters of extracellular fluid enter the lumen each day and are resorbed. The bulk of the accompanying calcium, however, is precipitated and excreted with the feces. Serum calcium and phosphate levels are also affected by renal excretion. Phosphate is resorbed in the proximal tubule but less efficiently than is Ca^{++}. Bone mineralization is disturbed in renal diseases either by an excessive loss of phosphate as in renal tubular diseases, or excessive retention when the glomerular filtration rate falls below 25 ml per minute. The hypocalcemia commonly associated with phosphate retention in chronic uremia involves several mechanisms. Prominent among them is the destruction of renal tissue needed for synthesis of $1,25\text{-}(OH)_2D_3$.

There probably is more to the story. The recent observation that prostaglandin E_2 can stimulate bone resorption and hypercalcemia in experimental conditions may add a new dimension to our understanding of the regulation and diseases of bone in the next few years.

EFFECTS OF OTHER HORMONES ON BONE

The development and remodeling of the skeleton also are profoundly influenced by hormones other than PTH and CT. Some hormones exert their action directly on the skeletal target; others have indirect effects, the net result of interactions with other endocrine glands.

The growth hormone (GH) of the anterior pituitary gland stimulates growth of many connective tissues, including the epiphyseal cartilage and periosteum. The levels of GH in the serum, as measured by radioimmunoassay, are very labile. Although this hormone is required for skeletal growth, there is little correlation between the fasting serum levels and the growth of normal children. GH is not itself the physiologically stimulatory agent but it is required for generating the hormone that is: somatomedin (sulfation factor). Somatomedin is formed in the liver under the influence of GH. Whether it is a derivative of GH or an entirely different molecule is unknown.

The sexual dimorphism of the skeleton and the cessation of linear growth at maturity result largely from the action of the gonadal hormones. Although androgens have a general anabolic effect and may thus cause a mild acceleration of bone growth, their principal action on the skeleton is to stimulate ossification and closure of the epiphyses. In the absence of thyroid hormone, GH is unable to permit epiphyseal ossification. In the absence of sex hormones, GH does not cause epiphyseal union.

In the established skeleton, adrenal glucocorticosteroids and thyroid hormone favor resorption of bone, while estrogens reduce it.

REPAIR OF BONE

When a bone is fractured, hemorrhage takes place between the broken ends, beneath the torn or stripped-off periosteum, and into the surrounding muscle. Organization of the hemorrhage is accomplished by ingrowth of a pluripotential granulation tissue from the adjacent bone marrow, periosteum, and muscle. From the granulation tissue emerges a variegated primordial skeletal tissue, the callus. Fibroblasts in it give way to chondrocytes and osteoblasts. The appearance in the callus varies with the location and stage of development. The callus external to the bone envelopes the defect and thus is called the ring callus. Proliferation of cartilage and osteoid tissue is particularly prominent near the periosteal cambium layer. The callus provides a provisional scaffold which ultimately is replaced by functionally adequate bone in accordance with Wolff's law. The speed at which these events take place depends on local and systemic factors. Osteoblasts are found within the callus in 10 days. Local factors in the further evolution include the magnitude of the defect and its immobilization. Although a minor degree of motion may be helpful, approximation and alignment of the broken ends and immobilization are required for proper healing. In their absence, and sometimes despite them, non-union may occur. A false joint (pseudarthrosis) may form in the cartilage of the callus if immobilization is inadequate (Fig. 2.6). Systemic factors include the nutritional state and the age of the individual. Repair is particularly rapid in growing children. Under proper mechanical and physiologic conditions, the provisional callus becomes fully ossified and remodeled until the original bone is more or

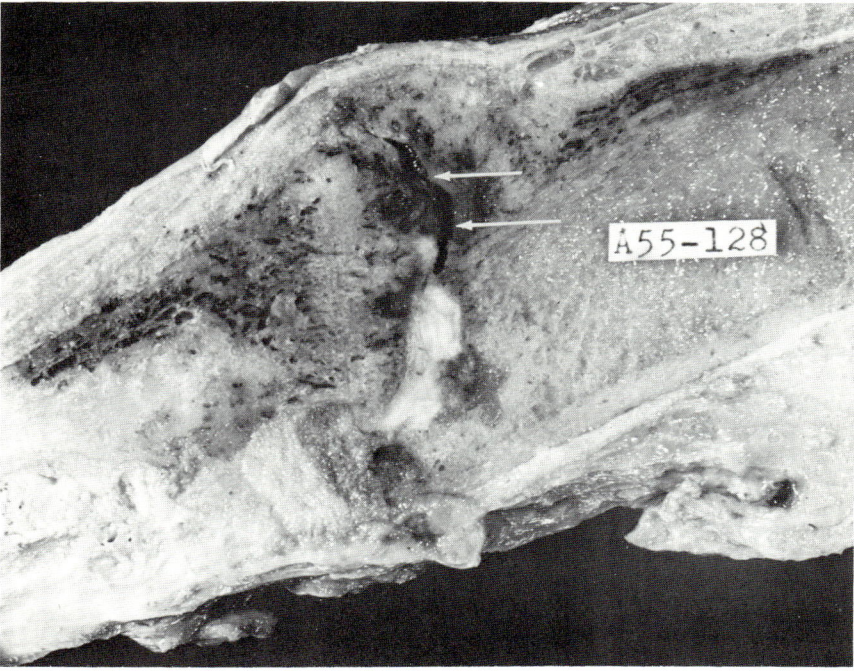

Figure 2.6. Pseudarthrosis, femur. The patient was a 65-year-old man whose lower extremities had been paralyzed by anterior poliomyelitis in infancy. The femur was fractured 10 years before death and never healed properly. The persistent jagged defect (*arrows*) is bordered on the upper edge by a bony ring callus and on the lower by white cartilage.

less perfectly recapitulated. Compact bone is more likely to form on the concave side of a deformity where compressive stresses predominate, and a thinner longitudinal cortex on the convex side where stresses are tensile.

EXTRAOSSEOUS MANIFESTATIONS OF DISTURBED BONE METABOLISM

Calcium balance studies require measurement of the quantities ingested, absorbed, and excreted into the feces and urine. The endogenous portion of the fecal calcium may be estimated from the appearance in the feces of radioactive calcium injected intravenously. In practice, however, the excretion of calcium is most often measured by the urinary loss; on a calcium-restricted diet the normal figure is 200 mg per 24 hours.

Hypercalcemia characteristically leads to fatigue and weakness by disturbing neuromuscular membranes. Vomiting also is prominent and death may result when serum calcium values exceed 18 mg per 100 ml. In protracted hypercalcemia, metastatic calcification, *i.e.*, deposition of calcium salts into soft tissues that were not the site of previous abnormality, may occur. Metastatic calcification is most often found in renal tubules, alveolar walls, gastric mucosa, and myocardium. Occasionally, it also is seen in joint tissues, *e.g.*, in hyperparathyroidism and hypervitaminosis D. The basis of the predilection for certain tissues is unknown; in the case of the viscera, local alkalinity may favor precipitation of the mineral.

Severe hypocalcemia (serum values below 7 mg per 100 mg) results in increased neuromuscular irritability manifested by tetany. Convulsive twitches of the face are early signs. This situation most often results from surgical extirpation of the parathyroid glands. In uremic hypocalcemia associated with renal osteodystrophy, tetany is infrequent because most of the Ca is in ionic form.

Small quantities of hydroxyproline-containing peptides arising from degradation of collagen pass into the urine. These quantities increase with extensive resorption of bone but the usefulness of this determination in metabolic bone disease is limited because other collagens contribute to the pool. Urinary hydroxyproline levels in bone disease generally parallel those of serum alkaline phosphatase. In growing children, values for the latter are 150 international units; in adults, 85.

REFERENCES

ANAST, C. S., AND CONAWAY, H. H. Calcitonin. Clin. Orthop. 84:207, 1972.

BONUCCI, E. The locus of initial calcification in cartilage and bone. Clin. Orthop. 78:108, 1971.

BROOKES, M. The Blood Supply of Bone. An Approach to Bone Biology. New York, Appleton-Century-Crofts, 1971.

DAUGHADAY, W. H. Sulfation factor regulation of skeletal growth. A stable mechanism dependent on intermittent growth hormone secretion. Am. J. Med. 50:277, 1971.

DELUCA, H. F. The kidney as an endocrine organ for the production of 1,25-dihydroxyvitamin D_3, a calcium-mobilizing hormone. N. Engl. J. Med. 289:359, 1973.

FOURMAN, P., AND ROYER, P. Calcium Metabolism and the Bone. 2nd ed. Oxford, Blackwell, 1968.

GLIMCHER, M. J. A basic architectural principle in the organization of mineralized tissues. Clin. Orthop. 61:16, 1968.

HARRIS, W. H., AND HEANEY, R. P. Skeletal Renewal and Metabolic Bone Disease. Boston, Little Brown, 1970.

IRVING, J. T. Calcium and Phosphorus Metabolism. New York, Academic Press, 1973.

JACKSON, W. P. U. Calcium Metabolism and Bone Disease. London, E. Arnold, 1967.

JOWSEY, J., et al. Quantitative microradiographic studies of normal and osteoporotic bone. J. Bone Joint Surg. 47A:785, 1965.

RAISZ, L. G. Physiologic and pharmacologic regulation of bone resorption. N. Engl. J. Med. 282:909, 1970.

SHIM, S. S., MOKKHAVESA, S., MCPHERSON, G. D. AND SCHWEIGEL, J. F. Bone and skeletal blood flow in man measured by a radioisotopic method. Can. J. Surg. 14:38, 1971.

TASHJIAN, A. H., JR., VOELKEL, E. F., GOLDHABER, P., AND LEVINE, L. Prostaglandins, calcium metabolism and cancer. Fed. Proc. 33:81, 1974.

ZIPKIN, I. (Ed.) Biological Mineralization. New York, J. Wiley & Sons, 1973.

3
JOINTS

Although joints may be classified many ways, only three are useful for present purposes: diarthroses, movable, cavity-containing joints; synarthroses, non-movable joints; and amphiarthroses which have limited mobility. The sutures of the skull are examples of synarthroses. The intervertebral discs are amphiarthroses and along with diarthroses are the seat of important musculoskeletal diseases.

DIARTHROSES
Articular Cartilage

The function of a diarthrosis is to be a pivot of the moving skeleton (Fig. 3.1). The motions are primarily oscillatory. Articular cartilage is the bearing surface and has unique physicochemical and biologic properties that keep it from wearing out over the years. The forces exerted on joint surfaces are very high as a result of leverage of the muscles acting about the pivot. In a deep knee squat, for example, a man weighing 150 pounds exerts a force of 1000 pounds on the patellofemoral cartilages. These pressures far exceed those in the blood stream and articular cartilage accordingly has no blood vessels. The avascularity is a mechanism allowing only slow turnover and stability of the tissue for its bearing function. In all physiologic and pathologic circumstances where cartilage acquires a blood supply, it becomes converted into bone or another type of connective tissue.

Articular cartilage is of hyaline rather than fibrous type except in the temporomandibular and sternoclavicular joints. It is bonded tightly to the bone through an intervening zone of calcification. The superficial edge of the calcified cartilage is the "tidemark," so-called because it denotes advancing waves of calcification. It is this structure which appears as the "joint surface" in x-ray films. The roentgenographic "joint space" does not correspond to the synovial cavity so much as to the non-calcified articular cartilage which is radiolucent. The surface of articular cartilage appears smooth to the naked eye but actually has fine undulations measuring up to 2 μ in depth. Although it has high tensile strength, cartilage is fairly compliant. As a result, loads applied to it are distributed over a broader area as the joint surfaces are compressed and

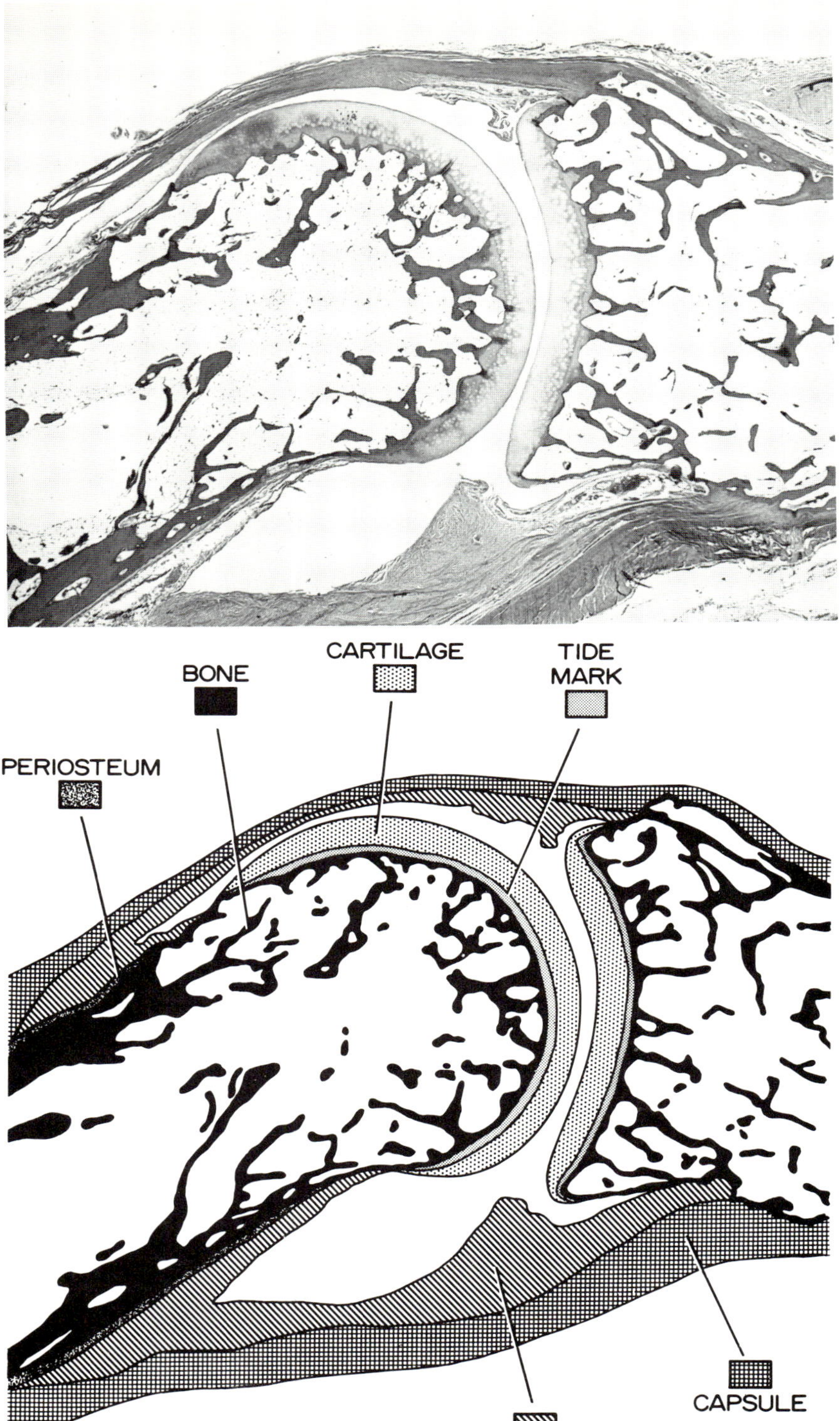

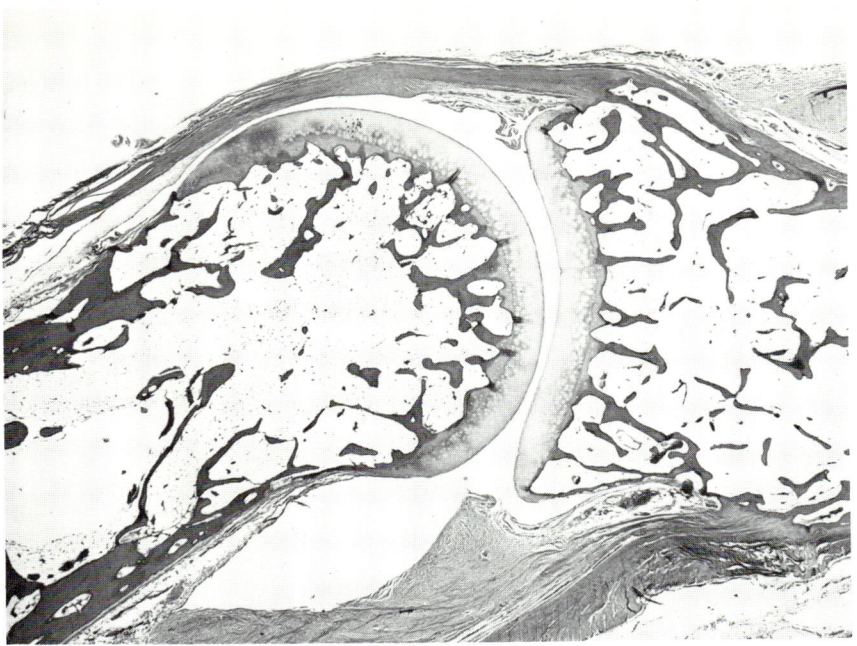

Figure 3.1. Normal interphalangeal joint.

the pressure (pressure = force/area) is reduced. These properties result from the high content of water (80 percent) of the cartilage. Of the remaining dry mass, approximately half, 10 percent of the total, is collagen, and half is ground substance.

The collagen is finely dispersed and woven into a three-dimensional network (Fig. 3.2). The fibrils are randomly arranged but their net orientation, as detected with polarized light, is in three zones. A tangential layer is arranged primarily parallel to the surface of the cartilage. A deep radial zone is oriented at a right angle to the surface. A transitional zone unites the preceding in arched arrangement or Benninghof arcades.

The principal mucopolysaccharides of the ground substance of articular cartilage are chondroitin sulfates and some keratan sulfate. Only a small quantity of hyaluronate is present, particularly at the surface of the joint. The oncotic pressure of the proteoglycans causes the ground substance to imbibe water and thereby inflate the collagen into a taut network. When a load is placed on the surface, as in weight-bearing, fluid is expressed from the matrix so long as the applied hydrostatic pressure exceeds the imbibition pressure of the cartilage. The water extruded into the joint space is available for lubrication of the joint ("weeping" lubrication). The process is reversed when the load is removed from the cartilage. This provides a mechanism for pumping nutrients through the avascular tissue. There is relatively little diffusion or nourishment from the subarticular bone marrow. Unlike collagen, the proteoglycans of articular cartilage undergo a modest turnover; in rabbits, the half-life is estimated as 8 to 60 days and in humans 250 days.

The cells of articular cartilage are sparse. They not only synthesize the proteoglycans but contain lysosomal enzymes (cathepsins) capable of digesting them. The chondrocytes have a predominantly glycolytic metabolism as an adaptation to their anaerobic environment. Consequently articular cartilage has a uniquely high lactate content. Although the cartilage does not regenerate readily following injury, the chondrocytes are capable of mitotic division when the pericellular matrix is dissolved.

Synovium

Joints arise embryologically through liquefaction of limb-bud mesenchyme. The non-cartilaginous lining of the movable joints is the *synovium* (Fig. 3.1). The term *synovia* applies to joint fluid and should not be confused with synovial tissue. There are several important biologic consequences of their origin from a pluripotential mesenchyme. The synovium is readily regenerated from granulation tissue following synovectomy. Despite the fact that several disorders in which the lining

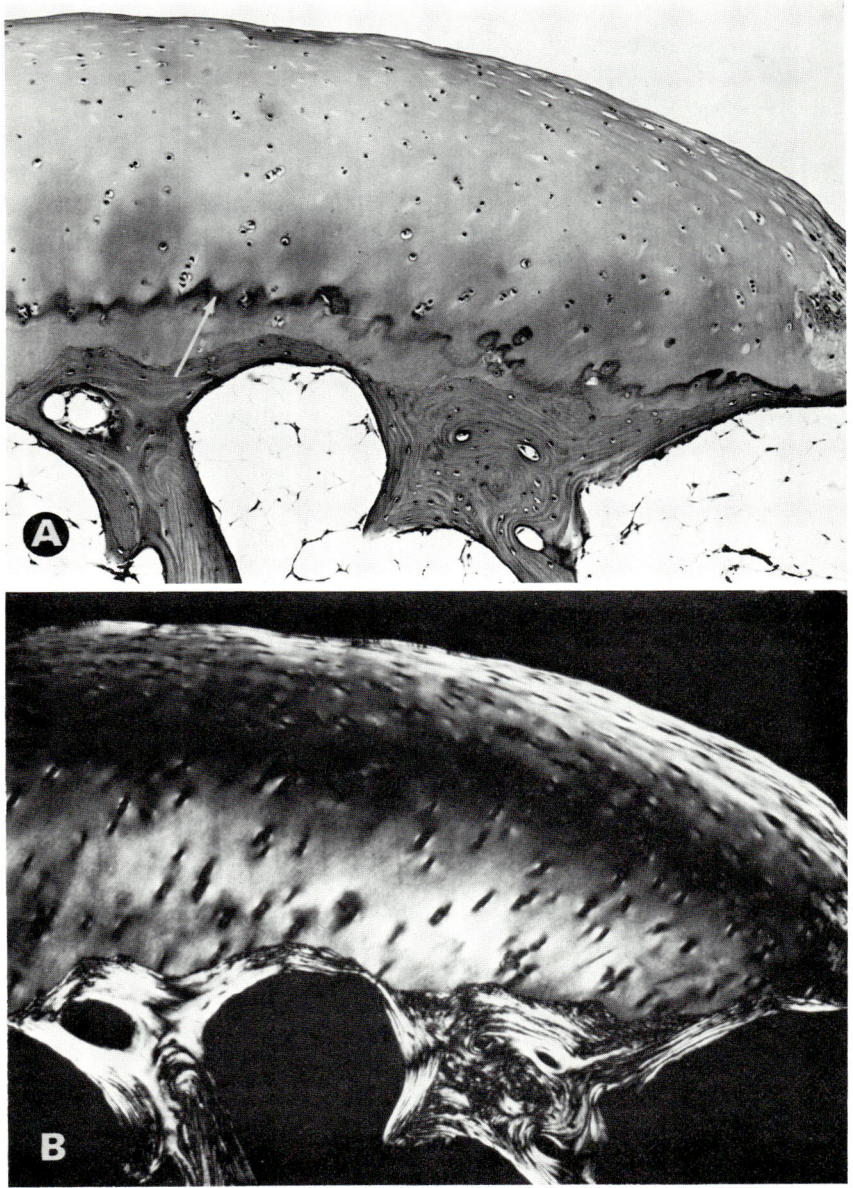

Figure 3.2. Normal joint surface. *A.* The articular cartilage is avascular and contains few cells. The *arrow* points to the "tidemark," the edge of the layer of calcified cartilage. *B.* The pattern of birefringence indicates the general orientation of collagen in the three zones: tangential, transitional, and radial. (Hematoxylin and eosin, ×72.)

of the body cavities (pericardium, pleura, peritoneum) as well as joints are inflamed are often called polyserositis, joints are not derivatives of coelomic epithelium and are therefore not serous structures or mesothelial. They have no basement membrane. The distinction between type A and B cells is not rigid; intermediate forms occur, particularly in disease states. The various tissues comprised by joints—synovium, fibroadipose tissue, fibrous and hyaline cartilage, ligaments, and periosteum—merge imperceptibly with each other. Metaplastic and neoplastic islands of cartilage and bone form within synovium at times, giving rise to synovial chondromatosis and osteochondromatosis. The configuration of joints is governed genetically but their full development *in utero* also requires appropriate mechanical forces. When fetal joints are unable to move, synovial cavities and auxiliary structures fail to develop. Thus, paralytic neuromuscular disorders in the embryo lead to a group of orthopedic diseases in infants, in which joints are incompletely formed and immobile (arthrogryposis multiplex congenita).

Synovial tissue is highly vascular, especially at the periphery of the articular cartilage where it is villous. The caliber of synovial capillaries is greater than those elsewhere in the body and many are fenestrated. They are coiled in glomeruloid fashion immediately beneath the synovial lining. These structural features favor a rapid exchange of water and small molecules between the blood and synovial fluid; indeed the joint cavity often is described as an expansion of the extracellular compartment of connective tissue. Synovial vessels anastomose extensively with those of the epiphysis and circulatory disturbances in one are commonly accompanied by changes in the other. Local changes in the synovial circulation attending use or injuries of joints lead to focal sclerosis of vessels and arteriovenous anastomoses. The inexperienced observer may confuse these with systemic vascular lesions.

Synovial Fluid

The synovial fluid is formed principally as a transudate from the vessels, but the synovial mucin is synthesized by the lining cells. The fluid resembles interstitial fluid elsewhere in the body. The concentration of small molecules is like that in serum except for small Donnan equilibrium shifts. The protein content is low, about one-third of that found in the serum. It consists principally of albumin but other proteins, including the complement system and enzymes, also are present. The most conspicuous feature of synovial fluid is its viscosity which derives from the mucin. Synovial mucin consists of hyaluronate and perhaps 1 percent of covalently bound protein. Both the amount (normally 350 mg per 100 ml) and degree of polymerization of the hyaluronate determine the viscosity. The viscosity is non-Newtonian: it varies with the rate at which the fluid is sheared. At high speeds of joint motion, the fluid thins out and hence is thixotropic. Synovial fluid strings out when a needle or loop is inserted into it and then slowly withdrawn. This property is related to but is not itself the viscosity of the mucin. Stringing out is often used as a first approximation of the quality of the mucin. Another related useful clinical procedure is the mucin clot test: when dilute (5 percent) acetic acid is added to normal synovial fluid, a firm clot is formed. When the mucin is defective, the clot is friable (Fig. 3.3).

Joints move with little friction. Even in the absence of synovial fluid, the coefficient of friction of articular cartilage is below that of an iceskate gliding on ice. The intrinsic slipperiness of cartilage results from the "weeping" properties of the matrix. Synovial fluid reduces the friction still further. The slimy quality of synovial fluid intuitively suggests that viscosity of the hyaluronate is instrumental in the lubrication, but recent evidence is to the contrary. Another molecule, perhaps a glycoprotein, may be the synovial lubricant.

Microscopic examination of synovial fluid is a valuable diagnostic tool. In counting cells, physiologic saline rather than the usual diluent solutions is employed because the latter are acidic and clot the mucin. Normally there are fewer than 300 white blood cells per mm^3 and no more than a quarter of these are polymorphonuclear leukocytes.

Figure 3.3. Mucin clot test showing (*left* to *right*) good mucin clot, fair, poor, and very poor.

Fine particulate matter is removed from joints by type A synovial lining cells. Larger aggregates appear in deep-lying macrophages and migrate through lymphatics. Regional adenopathy is therefore common in arthritic disorders.

Innervation

The innervation is involved in four principal manifestations of joint disease: pain, muscle spasms, reflex dystrophies, and the neuropathic arthropathies that complicate the loss of proprioception. Both pain and proprioceptive endings are located in capsule and ligaments. The synovium contains relatively few of these. Stretch is the principal stimulus to both sensibilities. Pain is usually poorly localized and may be referred to distant sites. The proprioceptive endings are of Ruffini and small lamellated corpuscle types. They perceive movement and position of the joint, and reflexly control posture and motion. By reflexly calling antagonistic muscles into play, they prevent the collagenous tissues from being overstretched and torn. There also are fine fibers in the adventitia of blood vessels which presumably serve vasomotor and pain-sensing functions.

AMPHIARTHROSES

The intervertebral discs have many analogies to articular cartilage. They differ from the latter basically in the separation of the ground substance and collagen into two relatively distinct compartments, the nucleus pulposus and anulus fibrosus, but the proteoglycans otherwise are similar. Adjacent to the vertebral bodies, the disc tissues blend with a hyaline cartilaginous plate. Blood vessels are absent, except for a few at the most peripheral portion of the anulus fibrosus, for the same reason that they are absent in articular cartilage: the pressures are too high within the disc to permit blood to flow. The collagen fibers are arranged in criss-crossing bundles so that the disc as a whole is moderately flexible even though the collagen is inextensible.

The high loads that are exerted on the spine are borne hydrostatically by the nucleus pulposus. The swelling properties of the ground substance are nowhere better seen than in the intervertebral disc removed at autopsy. In the freshly excised slab of vertebral column, the nucleus pulposus appears at a level with the adjacent tissues. Following immersion in water, the nucleus pulposus becomes greatly swollen, bulges, and appears gelatinous. The same phenomenon is observed even when the specimen is placed in formalin; the fixative hardens the protein but not the glycosaminoglycans which imbibe the water. The swelling is ordinarily constrained during life by the enveloping anulus fibrosus.

When rents develop in the anulus during life, swelling can proceed. The swollen nucleus pulposus may then be extruded through the defect and thereby give rise to the "herniated disc."

REFERENCES

CLARKE, I. C. The microevaluation of articular surface contours. Ann. Biomed. Eng. 1: 31, 1972.

COHEN, A. S. Synovial fluid. *in* Laboratory Diagnostic Procedures in the Rheumatic Diseases, p. 2, edited by A. S. Cohen. Boston, Little Brown, 1967.

DEE, R. Structure and function of hip joint innervation. Ann. R. Coll. Surg. Engl. 45:347, 1969.

DRACHMAN, D. B., AND SOKOLOFF, L. The role of movement in embryonic joint development. Dev. Biol. 14:401, 1966.

FREEMAN, M. A. R. (Ed.) Adult Articular Cartilage. New York, Grune & Stratton, 1973.

HAMERMAN, D., ROSENBERG, L. C., AND SCHUBERT, M. Diarthrodial joints revisited. J. Bone Joint Surg. 52A:725, 1970.

KEISER, H., AND SANDSON, J. Immunology of cartilage proteoglycan. Fed. Proc. 32:1474, 1973.

MCDEVITT, C. A. Biochemistry of articular cartilage. Nature of proteoglycans and collagen of articular cartilage and their role in ageing and in osteoarthrosis. Ann. Rheum. Dis. 32:364, 1973.

MORRISON, R. I. G., BARRETT, A. J., DINGLE, J. T., AND Prior, D. Cathepsins B_1 and D action on human cartilage proteoglycans. Biochim. Biophys. Acta 302:411, 1973.

OGSTON, A. G., AND WELLS, J. D. The osmotic properties of sulphoethyl-sephadex. A model for cartilage. Biochem. J. 128:685, 1972.

ORONSKY, A., IGNARRO, L., AND PERPER, R. Release of cartilage mucopolysaccharide-degrading neutral protease from human leucocytes. J. Exp. Med. 138:461, 1973.

RADIN, E. L., AND PAUL, I. L. A consolidated concept of joint lubrication. J. Bone Joint Surg. 54A: 607, 1972.

SOKOLOFF, L., MALEMUD. C. J., SRIVASTAVA, V. M. L., AND MORGAN, W. D. *In vitro* culture of articular chondrocytes, Fed. Proc. 32:1499, 1973.

SWANN, D. A., AND RADIN, E. L. The molecular basis of articular lubrication. 1. Purification and properties of a lubricating fraction from bovine synovial fluid. J. Biol. Chem. 247:8069, 1972.

TUSHAN, F. S., RODNAN, G. P., ALTMAN, M., AND ROBIN, E. D. Anaerobic glycolysis and lactate dehydrogenase (LDH) isoenzymes in articular cartilage. J. Lab. Clin. Med. 73:649, 1969.

4

SKELETAL MUSCLE

Skeletal muscle constitutes 40 to 50 percent of the normal body weight. Although the fibers originate independently of a nerve supply, their ultimate maintenance and function are profoundly influenced by their innervation. In practice, it frequently is difficult to determine whether a disease arises in the muscle proper or in its nerve supply. For this reason, we must deal with the entire motor unit, *i.e.*, the lower motor (α) neuron as well as the neuromuscular junction and myofibers that it supplies. Each of these structures shares common physiologic features: initiation of their action requires electrical depolarization of the cell membrane; furthermore, the subsequent accomplishment of their specific functions requires ionic calcium.

During rest, the interior of the cells is electronegative with respect to the exterior. The potential difference across the cell membrane arises from differences in the concentrations of potassium ions within and external to the cells. Intracellular K^+ concentrations are 20 to 50 times higher than in the extracellular fluid, while Na^+ is 10 times more concentrated in the latter. An active metabolic pump serves to expel Na^+ from the cell and permits K^+ to re-enter it. The potential difference across the cell membrane is just sufficient to keep the ions from diffusing across their concentration gradients. The cell membrane is thus polarized. When the membrane no longer separates the charges, it is depolarized. Transmission of the depolarization along the surface of the cell membrane gives rise to the action potential. Electromyography, which analyzes these potentials, is an important tool in clinical as well as investigative aspects of neuromuscular diseases. It follows, too, that electrolyte disturbances, *e.g.*, of potassium and those of calcium attending bone disease, cause serious dysfunctions of skeletal muscle.

MUSCLE CELLS

Muscle cells (myofibers) have characteristic cross-striations and are multinucleated. Unlike cardiac muscle, the nuclei of skeletal myofibers are located at the periphery of the cell rather than the center.

Furthermore, skeletal muscle fibers are not separated by intercalated discs and do not branch. They are very long; some at least extend uninterrupted from origin to insertion as much as 30 cm. The myofibers are individually encased in a fine reticular connective tissue sheath, the endomysium. The endomysium coalesces with sheaths surrounding the muscle bundles (perimysium) and the entire muscle (epimysium) which converge to form the tendons. The connective tissue sheaths thus constitute a network for collecting and communicating muscle forces to the bones.

Muscle fibers are derived from mesoderm. They differentiate from the mesenchyme that forms the skeleton (sclerotome) and the skin (dermatome) without neural influences. From the primordial myotome cells that are capable of mitosis, emerges a series of cells (myoblasts) that can synthesize contractile proteins but no longer can divide. The myoblasts then coalesce to form primitive syncytia, the myotubes, in which the nuclei are centrally placed and myofibrils are sparse. Finally maturation proceeds so that the myofibrils occupy the center of the cells and the nuclei are displaced to the periphery where they lie beneath the plasma membrane (sarcolemma). It is presently believed that sarcolemmal nuclei cannot undergo mitotic division. If this is so, the muscle proteins undergo a rapid turnover but the cell proper persists throughout the life span of the individual. Muscle grows in length by addition of new sarcomeres to either end of the fibers. The remarkable ability of muscle to hypertrophy comes about through an increase in the number of myofibrils within the cells rather than a proliferation of the myofibers.

Aside from their nuclei and plasma membrane, muscle cells have four principal compartments: the myofibrils which are the contractile elements; the sarcoplasm, the fluid matrix of the cytoplasm; the sarcoplasmic reticulum, analogous to the endoplasmic reticulum of other cells; and the T-system, a series of transverse tubular invaginations of the plasma membrane which carry electrical charges from the surface of the cell to the myofibrils in the interior. The striated appearance of skeletal muscle arises from alternating stacks of actin and myosin within the myofibrils. When examined with a polarizing microscope, anisotropic (A) and isotropic (I) bands are seen. Each major repeating period (sarcomere) is delimited by a non-contractile disc, the Z-line, which is composed of the protein tropomyosin. The A band corresponds to the position of the myosin molecule.

Actin and myosin are filamentous proteins. They are oriented longitudinally and partly overlap with each other within the sarcomere (Fig. 4.1). In the aggregate, they constitute the myofibrils. Contraction is accomplished by a chemical linkage between the two so that actomyosin

Skeletal Muscle   45

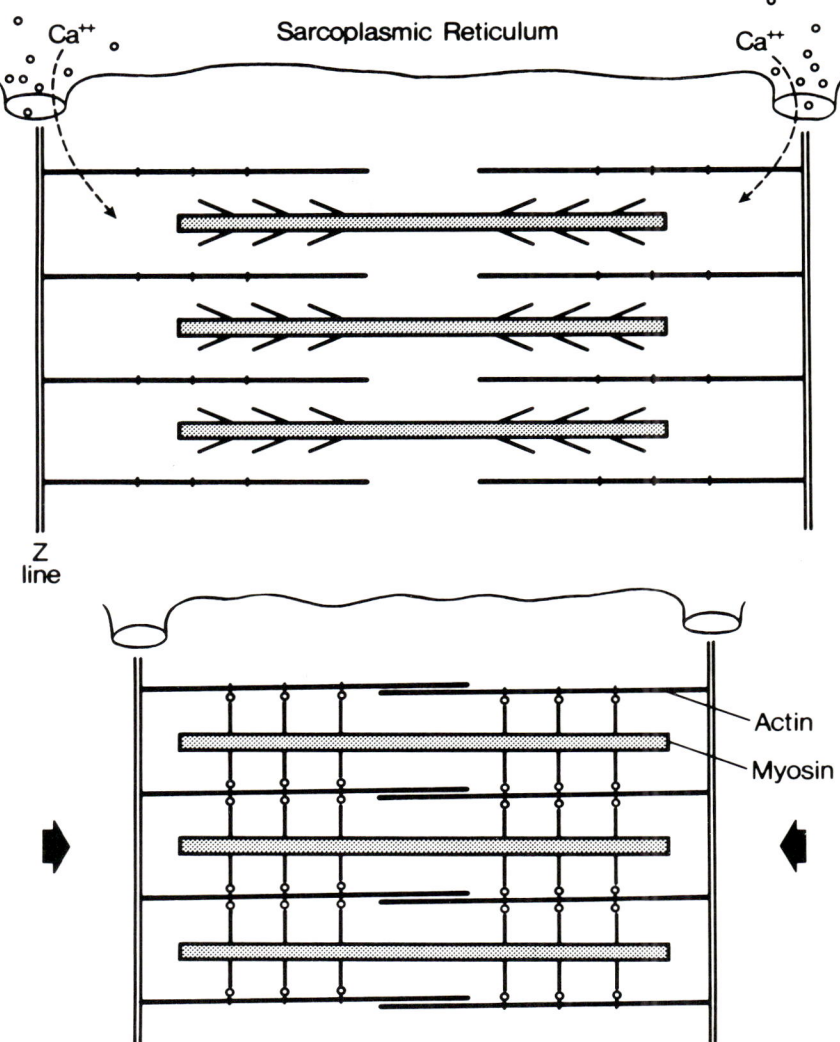

Figure 4.1. In the contraction of the sarcomere, filaments of actin are drawn over the myosin by a chemical linkage between the two molecules, and may actually overlap with each other in extreme shortening. The reaction is mediated by Ca^{++} released from the sarcoplasmic reticulum.

is formed. Cessation of contraction comes about when the linkage is broken. The actin filaments are attached to the Z-lines. All movement of muscle, active or passive, is brought about by sliding of the actin over the myosin filaments thereby drawing the Z-lines closer toward each other. A globular end piece of the myosin molecule has a calcium-stimulated adenosine triphosphatase (ATPase) activity. Adenosine triphosphate (ATP) is also a component of the myosin, and adenosine diphosphate (ADP) of actin. The union between the two molecules is mediated by Ca^{++} which makes chelate bonds between these ADP and ATP moieties. Troponin, a protein attached to the actin, has a high affinity for Ca^{++} and must play an important role in regulating the action of this ion in the contractile process. The Ca^{++} at rest lies within the sarcoplasmic reticulum. The action potential in some manner releases Ca^{++} from the latter to the myofibrils so that contraction can occur; thereafter it returns to the sarcoplasmic reticulum. The ability of the latter to reaccumulate Ca^{++} is sometimes referred to as "relaxation factor." Another ATPase, magnesium dependent, is involved in pumping the Ca^{++} back into the lateral cisternae of the sarcoplasmic reticulum. The rapid coupling of excitation and contraction is made possible by the spatial proximity of the T-system to the sarcoplasmic reticulum and the junction of the I and A bands of the myofibrils.

The several components of the energy-producing system, discussed in the next section, are packaged in separate compartments. The phosphorylases and other enzymes of the glycolytic cycle, as well as the glycogen, are located within the aqueous sarcoplasm. The Krebs cycle enzymes and cytochrome oxidase are mitochondrial. The histochemical partition of these elements permits recognition of the two groups of muscle fibers that cannot be distinguished in conventional paraffin sections (Fig. 4.2). In some species, red fibers can be distinguished from white ones because they occur in different muscle groups, e.g., in the turkey thigh vs. breast. In man they are more intimately mixed and occur in roughly equal proportion among the various muscles. The red (Type I) fibers derive their color from their content of myoglobin and perhaps also from greater vascularity. They are identified histochemically by their high mitochondrial content and therefore from their staining for cytochrome oxidase and succinic dehydrogenase. The Type II, white fibers, by contrast, have less stainable mitochondrial enzymes, but a more prominent sarcoplasmic phosphorylase and glycogen content. Glucose and fatty acids presumably are the principal metabolic substrates of the former; and glycogen of the latter. Type I fibers are of slow twitch and Type II of intermediate fast twitch variety. Although both types of fiber occur within a single fascicle, each is supplied by a different lower motor neuron. These histochemical characteristics are widely employed in biopsies for the diagnosis of muscle disease. They have suggested new insights into the innervation and regeneration of muscle and, in the judgment of some experts, are shaking long-taught concepts of these disorders.

Figure 4.2. Normal myofiber types, myosin ATPase stain. The Type II fibers are the more darkly stained. (×200.)

BIOENERGETICS OF MUSCLE

The mechanical efficiency of muscle (the proportion of energy actually converted into work against a load) is about 15 percent. This figure is comparable to that of the more highly efficient man-made machines, such as steam turbines. The remainder of the energy appears as heat generated primarily during the contraction process. Energy also is stored elastically in the connective tissue sheaths and tendons when muscles are stretched beyond their resting length.

The immediate source of energy for muscle is the hydrolysis of adenosine triphosphate (ATP) to adenosine diphosphate (ADP), but ultimately it derives from foodstuffs that have been converted into glycogen. These pathways and some of their clinically useful enzymes are summarized in a much oversimplified scheme in Figure 4.3. The hydrolysis of ATP and ADP is catalyzed by the myofibrillar ATPase and the reaction takes place almost instantaneously. Unbound ATP in the sarcoplasm acts as energy source and participates in the calcium pump. The ATP is present in relatively small quantities and is repeatedly regenerated by rephosphorylation of ADP, phosphate being transferred

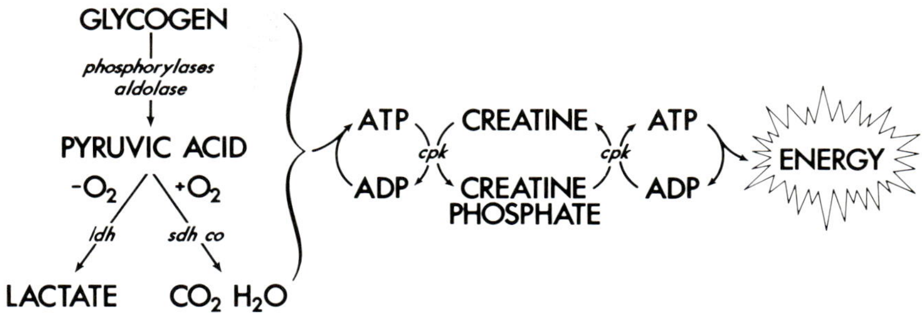

Figure 4.3. Simplified representation of energy pathways of muscle indicating enzymes commonly employed in clinical diagnosis or histochemistry of muscle. Abbreviations: *ldh*, lactate dehydrogenase; *sdh*, succinate dehydrogenase; *co*, cytochrome oxidase; *cpk*, creatine phosphokinase.

to the latter from creatine phosphate. Creatine phosphate is a store of high energy phosphate characteristic of muscle but cannot itself supply energy for muscular contraction as ATP does. The enzyme involved in the transfer of phosphate from creatine phosphate to ATP is creatine phosphokinase (CPK).

Glycolysis is the partial catabolic process by which glycogen is oxidized to pyruvic acid. It does not require oxygen. Glycogen and the glycolytic enzymes, including glycogen phosphorylase, are located in the sarcoplasm. The further fate of the pyruvate depends on the availability of oxygen. In the absence of oxygen, pyruvate is reduced to lactate by lactate dehydrogenase (LDH) with the coenzyme NADH. There are five related molecular forms of LDH, their chemical and functional differences depending on the proportion of the two subunit proteins, M and H. M is the isozyme characteristic of skeletal muscle and suitable for anaerobic glycolysis. H is characteristic of heart muscle and channels pyruvate more to aerobic than anaerobic pathways. When oxygen is available, oxidation of pyruvate can be carried to completion, the end product being CO_2 and H_2O and the yield of ATP accordingly much larger. The oxidative pathway for this is the tricarboxylic acid cycle and it is located in the mitochondria. The terminal catalyst is cytochrome oxidase, also located in the mitochondria. Muscle also contains the oxygen-storing heme protein, myoglobin. Only small quantities are present in man but in the muscle of diving species such as whales and penguins, which must store oxygen for long periods, the content of myoglobin is enormous. Myoglobin has some similarities to, but its protein is distinct from, that of hemoglobin.

Leakage of enzymes from damaged myofibers into the serum permits valuable measurements for diagnosing and following the course of muscle diseases. In general elevated serum levels of muscle enzymes are characteristic of myopathic rather than neuropathic disorders. Two aminotransferases, serum glutamic-oxalacetic (SGOT) and glutamic-pyruvic (SGPT), often used for this purpose, are part of protein synthesis as well as energy production by the cells. Serum CPK levels have a greater degree of specificity for muscle damage than most of the other enzymes because CPK is not elevated in liver disease. It is, however, high following myocardial infarction, childbirth, and excessive exercise. The skeletal muscles contain more than 95 percent of the body's creatine. The creatine is synthesized by the liver and transported to the muscle where it is catabolized into creatinine. Creatinine is diffusible and excreted into urine, normally at the level of 1.2 to 1.8 g per day. The corresponding figures for creatine are 60 to 180 mg. The urinary ratio of creatine:creatinine is disturbed in certain muscle diseases.

THE MOTOR NERVE FIBER

A single motor neuraxon branches at its terminus and sends a twig to each myofiber in its group; there may be between 5 and 200 of these. The units are not all activated at the same time but are integrated to work in relays, yielding a smooth contraction and minimizing fatigue. The neuraxon contains fine neurofibrils in addition to other usual plasma membrane and cytoplasmic structures. There is an active turnover of neurofibrillar and other cell proteins within the axon. Inasmuch as this requires nuclear synthetic mechanisms, there is in effect a continuous formation of cytoplasm which flows down the axon at a rate of about 1 mm per day.

The electrical properties of the cell are active and passive. For the action current to be propagated along the axon, the longitudinal resistance of its cytoplasm must not be excessive. At the same time, the transverse resistance must be high so that the current will not leak out across the membrane and the signal be dissipated. This transverse resistance is accomplished through insulation by the myelin sheath. Neuraxons are invested by a layer of neurilemmal (Schwann) cells that are distinct from the endoneurial connective tissue sheath. Schwann cells rotate and spin a concentric wrapping of plasma membrane about the neuraxons during their development. Myelin is the residual laminated lipoprotein of the Schwann cell membrane. The motor nerve fibers of muscle are thick and heavily myelinated. In a large myelinated fiber there are at least 100 such rotations. Each rotation reduces the transmembrane potential by half, so myelin is a good insulator. It is only at discontinuities in the myelin sheath, that an action potential can be regenerated and serve as an active relay for propagation of the signal.

THE NEUROMUSCULAR JUNCTION

The terminal portion of the neuraxon divides into fine filaments, each of which forms a single specialized junction with a muscle fiber: the motor end plate. The end plates are not distributed uniformly throughout muscle, but tend to be centrally located and excite the muscle toward both ends. They thus can be seen only in certain locations rather than in casual ones. In conventional histologic preparations they may be difficult to recognize. They appear as small aggregates of nuclei external to a muscle fiber, and the uninitiated may confuse them with infiltrating lymphocytes and mononuclear cells. The nuclei are in fact of Schwann cell and sarcolemmal origin. For biopsy studies, intravital staining with methylene blue is useful in identifying the axon terminals.

The axon terminal plugs into a synaptic trough in the sarcolemma of the motor end plate. A gap, 500 to 1000 Å wide, separates the two plasma membranes. The synaptic gap is sufficiently wide to prevent transmission of the signal from nerve to muscle except when a switching mechanism operates. Within the axon terminal are large numbers of vesicles (presynaptic vesicles) which contain the neurotransmitter, acetylcholine. When the axon terminal is depolarized, acetylcholine is discharged from the presynaptic vesicles into the synaptic gap. This mechanism requires ionic calcium. Acetylcholine is a cation and effectively depolarizes the postsynaptic surface, initiating muscle excitation. On the postsynaptic surface, the enzyme cholinesterase is located. It hydrolyzes the acetylcholine and thereby switches off the excitation. Cholinesterase is readily demonstrated at the motor end plate by histochemical means. Curare competes chemically with acetylcholine for postsynaptic receptors and thus is widely used as a muscle relaxant. Analogues of acetylcholine, such as succinylcholine and quaternary ammonium bases, also depolarize the postsynaptic surface. Their muscle-relaxing action results from their resistance to hydrolysis by cholinesterase; the surface of the muscle remains depolarized and cannot be re-excited. The parasympathomimetic agents, prostigmine and its analogues, are potent inhibitors of cholinesterase and are employed to encourage muscle contraction.

THE MUSCLE SPINDLE

Muscle spindles are encapsulated sensory organs which serve as strain gauges in skeletal muscle (Fig. 4.4). They are fusiform and measure up to 5 mm in length. Their number varies but they are roughly $\frac{1}{10}$ as many as nerve fibers. The capsule is filled with fluid. It also contains 2 to 12 longitudinal modified muscle fibers, the intrafusal fibers. Intrafusal

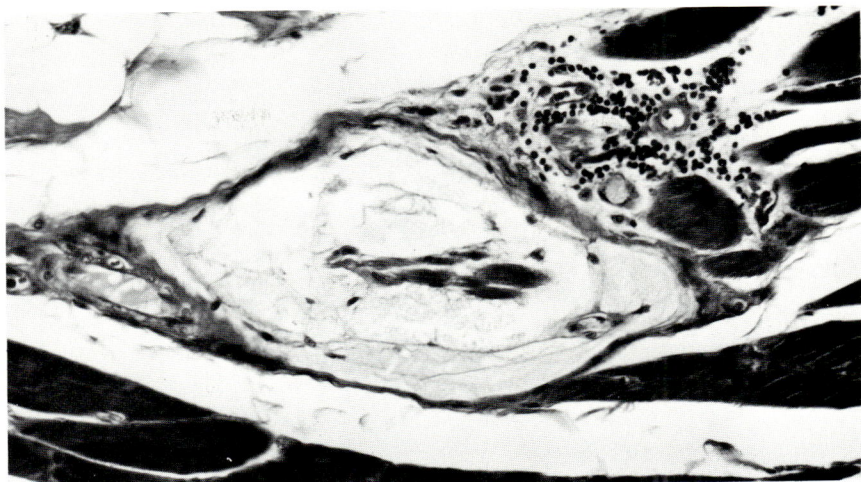

Figure 4.4. Normal muscle spindle in a case of myasthenia gravis. The fibers in the center are intrafusal muscle; nerve endings cannot be seen without special stains. The aggregate of lymphocytes about minute vessels above and to the right of the spindle is a "lymphorrhage" and is part of the myasthenic lesion. (Hematoxylin and eosin, ×225.)

fibers have tapered ends. The equatorial part is devoid of myofibrils and contains many nuclei. Extending from this region the cells become myotubular and, toward their ends, cross-striated. Afferent nerve endings encircle the equatorial portion of the intrafusal fibers and are stimulated when the latter are stretched. The frequency of the discharge is proportional to the velocity and magnitude of the stretch. The afferents communicate through interneurons in the spinal cord with α and γ motoneurons. The latter are small neurons that supply efferent fibers to the intrafusal fibers. They thus, through a closed-loop servosystem, maintain normal reflex control of synergistic and antagonistic muscle groups in postural and phasic activities.

There also are stretch receptors in tendons. The muscle spindle is not involved in sensation of position and movement. Receptors for the latter are located in the joint capsule and periarticular tissues.

REPAIR OF MUSCLE

Skeletal muscle has limited capacities to regenerate itself. Following trauma or surgical interruption, the pathologist commonly finds replacement by scar tissue rather than new muscle. It has already been noted that muscle cells are incapable of mitotic division once they become multinucleated. Nevertheless there are reasons for believing that some repair is possible. This is true particularly when the endomysial sheath is not disrupted. Following segmental necrosis of the myofiber, clusters of large vesicular sarcolemmal nuclei having prominent nucleoli may appear (see Fig. 5.1). The adjacent sarcoplasm contains a few myofibrils but is stained gray-blue by hematoxylin. The latter staining property results from large quantities of ribosomal RNA, a feature of many blastic cells. This is the basis for regarding such cells as regenerating rather than degenerating. This evidence of regeneration was once regarded by many pathologists as a useful diagnostic feature in distinguishing inflammatory and certain other muscle diseases from muscular dystrophies. Opinion is divided as to whether mononucleate cells, from which reparative myogenesis proceeds after injury, are derived from breaking up and dedifferentiation of the injured myofiber or from primordial myoblasts that persist from fetal into adult life. Such resting or satellite cells are said to lie next to myofibers beneath the endomysial sheath but are not readily recognized histologically in man.

To the extent that muscle fibers regenerate, they do so without benefit of a nerve supply. In time, however, new axonal fibers proliferate and migrate to the regenerating myofibers, creating new myoneural junctions. The differentiation of myofibers into the two types is determined always by the lower motor neuron that resupplies them. This presumably is accomplished by trophic factors that the neurons synthesize and secrete. The reinnervating fibers may or may not come from the original neuron. If they do not, they impart a new differentiation to the regenerating muscle fibers. This is a mechanism allowing changes in the population of fiber types in various muscle diseases.

Following several sorts of injury, *e.g.*, surgery, irradiation, and trichinal infestation, multinucleated giant cells form from skeletal muscle. The myogenic giant cells may completely lack myofibrils and simulate foreign body or Langhans cells.

ELECTROMYOGRAPHY (EMG)

In clinical practice, the electrical activity of muscle is studied by inserting fine needle electrodes directly into the muscle. The resulting signal is displayed visually using an oscilloscope or other recorder, and acoustically with a loud speaker. The parameters evaluated are the amplitude, duration, frequency, and configuration of the waves. Measurements are made under three conditions: rest, minimal voluntary contraction sufficient to detect only a single motor unit, and maximal voluntary contraction (Fig. 4.5).

The injury caused by introduction of the needle causes a brief burst of insertional activity. It is followed in normal resting muscle by electrical silence (Fig. 4.5a). Thus muscle has no tonic contraction. The "tone" of

Skeletal Muscle   53

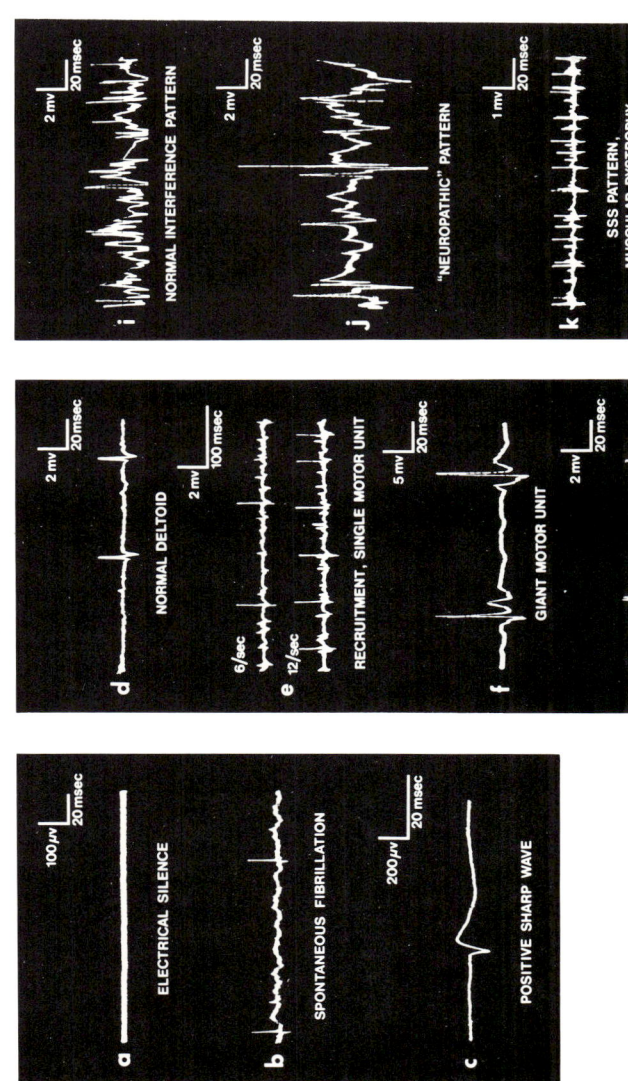

Figure 4.5. Basic electromyographic patterns. See text.

muscle at rest resides in the viscoelastic properties of its connective tissue components and myofibers.

The duration of the normal spike usually is 5 to 12 milliseconds (thousandths of a second) and the amplitude is of the order of 0.5 millivolts (Fig. 4.5d). The amplitude is a function of the number of myofibers in the motor unit that are being excited. The frequency of the firing increases with the degree of muscular effort (Fig. 4.5e) and additional motor units may be recruited as well. Each motor unit spike has a constant appearance at any given position of the sensing electrode and is distinguished from additional motor units by their own appearances and frequencies. The smallest number of spikes produced by a minimal contraction is called the recruitment frequency of that muscle. The maximal rate of activation of a motor unit is 50 per second, the intervals between them being the refractory period of the fibers. The overlay of spikes at high frequencies with recruitment of multiple motor units normally leads to complete interference patterns (Fig. 4.5i).

EMG is a generally useful but not an absolute means for determining whether a muscle disease originates within the myofibers proper or is secondary to a lesion in its nerve supply.

Denervation necessarily reduces the number of motor units during voluntary contraction. Approximately 2 to 3 weeks after its nerve supply is interrupted, muscle becomes highly irritable. Three useful EMG changes are associated with this: (1) Fibrillations, spontaneous contractions of isolated myofibers, occur and appear electrically as low, short, and infrequent spikes under conditions of rest (Fig. 4.5b). Fasciculation, a coarser and more visible spontaneous contraction of muscle than fibrillation, entails discharge of motor units rather than myofibers. It is most characteristic of amyotrophic lateral sclerosis and other anterior horn cell disease, but also occurs at times in peripheral neuropathies and even fatigue. (2) Insertional activity is increased. (3) Fortuitous penetration of the sensing needle into a damaged myofiber causes a sharp positive (downward) first wave (Fig. 4.5c). Partial denervation or other loss of a portion of the myofibers in a motor unit results in polyphasic spikes (those having more than 2 to 3 waves) because fewer individual fiber potentials are available to summate into the smoother wave form.

Reinnervation of muscle produces another brand of EMG abnormalities. The explanation for these has been made possible only recently by the histochemical differentiation of fiber types. They reveal that muscle fibers may be reinnervated in two ways: from the original or from a collateral axon. When the insult does not cause extensive disruption of the nerve, *e.g.*, in crush injury, the original axon or its twigs regenerate and the sprouts grown down their own endoneurial tube to the muscle.

The growth rate and conductivity of these sprouts are variable and they thereby cause the spike to become prolonged and more polyphasic (Fig. 4.5g). This is called the nascent motor unit. In collateral reinnervation, the nerve sprouts grow from an adjacent healthy axon which must therefore supply an even larger number of myofibers. It therefore results in an increased amplitude of the spikes (giant motor units, Fig. 4.5f). Thus the classic "neuropathic" EMG under maximal voluntary contraction is characterized by decreased motor units (incomplete interference patterns), high amplitude, and long-lasting spikes (Fig. 4.5j).

Unlike the preceding, damage to individual myofibers must reduce the amplitude of the spikes, and, if sufficiently extensive, reduce their duration. If the entire motor unit is not knocked out, the frequency is not reduced and commonly is increased as more and more recruitment is required for the muscle to accomplish its work. This pattern has long been called myopathic but may also theoretically result from fractional damage to nerve twigs as well. For this reason, many electromyographers now prefer to abandon the term "myopathic" and use instead descriptive words without etiologic implications.

The only pathologic state in which the EMG pattern is pathognomonic is myotonia. Here high frequency discharges of undulating amplitude follow a muscle contraction (Fig. 4.6). The most frequently affected sites are the thenar, forearm, and deltoid muscles.

EMG equipment may also be used to measure the conduction velocity of peripheral nerves. Using surface electrodes, two points on a motor nerve are stimulated in sequence across the skin while recording the action potentials of the muscle. Dividing the differences between the latencies (intervals between the stimulation and the muscle spikes) of the two points on the nerve by the distance between them yields the motor nerve conduction velocity. This varies from nerve to nerve and individual to individual and normally is in the range of 46 to 66 m per second. These values are reduced in demyelinating lesions of the axons and no conduction is found when both the axon and the myelin sheath are destroyed. Sensory nerve conduction velocities are more sensitive indicators of peripheral neuropathy than are the motor nerve values. Muscle conduction velocity is much slower than that of nerves, 1.3 to 4.7 m per second. There is a disparity between the time that elapses between stimulation of the distal end of the nerve and the time actually required for activation of the muscle. This "residual latency" measures primarily the synaptic delay at the myoneural junction and normally is 1.52 msec.

Muscle may be stimulated to contract following complete denervation only if stronger currents are employed. This procedure continues to enjoy some usage in clinical evaluation of the reinnervation of muscle.

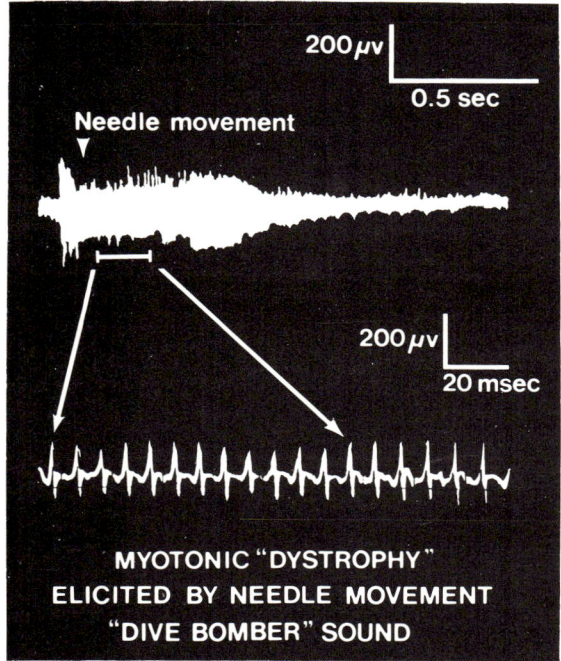

Figure 4.6. The protracted and undulating excitation following the slight stimulation is characteristic of myotonia.

Alternating (faradic) current is less effective than direct (galvanic) because the stimulus is constantly interrupted and the individual pulses of electricity short-lived. The disparity between the response to faradic and galvanic stimulation is called the reaction of degeneration.

MUSCLE BIOPSY

No single modality of examination is adequate for diagnosing or classifying all muscle diseases. In some, the clinical patterns, laboratory data, and EMG are sufficient. In others specific metabolic defects require biochemical analyses. Muscle biopsy is a valuable diagnostic procedure in still others but it entails certain specialized techniques.

The site for biopsy should not be so severely affected that the pathologic changes have progressed beyond a recognizable stage. The procedure is carried out under local anesthesia. Surgical trauma and certain chemical fixatives may cause the muscle to go into tetanic spasm and thereby become greatly distorted: the sarcoplasm appears hyalinized and pulled away from the endomysial sheath. It therefore is good practice to grasp the muscle segment with an isometric clamp and then excise the tissue, clamp and all. Disposable clamps for this purpose are available commercially. A portion of the specimen may be processed for conventional histologic examination but in many conditions it is desirable to freeze part for histochemical demonstration of relevant enzyme activities. A needle biopsy technique, recently described, may provide adequate tissue for certain purposes while avoiding unsightly scars.

Several histologic forms of myofiber degeneration are described at this point because they are non-specific and recur repeatedly in the chapters that follow. In each, the cells lose their striated appearance, or at least segments of them do. The sarcoplasm may appear homogeneous and oxyphilic; hence this degeneration is hyaline. When the sarcoplasm is coarsely fragmented, the degeneration is described as floccular. Less often the degenerated sarcoplasm is granular or cloudy. These various changes may coexist and have a similar significance. Vacuolar degeneration is less common and occurs in a narrower spectrum of diseases than the preceding. Several mechanisms can be distinguished ultrastructurally. Most of the vacuoles are membrane bound; lipid droplets are not. Basophilic degeneration corresponds to what we have already called regeneration of striated muscle. The histologic hallmark of necrosis is disappearance of sarcolemmal nuclei. More specific patterns of myofiber change are considered subsequently.

REFERENCES

AIDLEY, D. J. The Physiology of Excitable Cells. Cambridge, Cambridge University Press, 1971.
BOWMAN, J. P. The Muscle Spindle and Neural Control of the Tongue. Implications for Speech. Springfield, Ill., Charles C Thomas, 1971.
CARLSON, B. M. Cell and tissue interactions in regenerating muscle. *In* Muscle Biology, Vol. 1, p. 13, edited by R. G. Cassens. New York, Marcel Dekker, 1972.
ENGEL, W. K., AND WARMOLTS, J. R. The motor units; diseases affecting it *in toto* or in *portio*. *In* New Developments in Electromyography and Clinical Neurophysiology, p. 141, edited by J. R. Desmedt. Basel, S. Karger, 1973.
GOLDBERG, A. L., JABLECKI, C., AND LI, J. B. Effects of use and disuse on amino acid transport and protein turnover in muscle. Ann. N.Y. Acad. Sci. 228:190, 1974.
GOODGOLD, J., AND EBERSTEIN, A. Electrodiagnosis of Neuromuscular Diseases. Baltimore, Williams & Wilkins, 1972.
KAGEN, L. J. Myoglobin: Biochemical, Physiological and Clinical Aspects. New York, Columbia University Press, 1973.
McCOMAS, A. J., SICA, R. E. P., AND CAMPBELL, M. J. "Sick" motoneurones. A unifying concept of muscle disease. Lancet 1:321, 1971.
PODOLSKY, R. J. (Ed.) Contractility of Muscle Cells and Related Processes. Englewood Cliffs, N.J., Prentice-Hall, 1971.
SLOPER, J. C., BATESON, R. B., HINDLE, D., AND WARREN, J. Muscle regeneration in man and in mouse. *In* Regeneration of Striated Muscle and Myogenesis, p. 157, edited by A. Mauro, S. A. Shafiq, and A. T. Milhorat. Amsterdam, Excerpta Medica, 1970.

5

THE PAINFUL MUSCLE

Although somewhat artificial, it is useful to separate muscle disorders in which some sort of pain is a prominent feature from those in which it is not. Pain frequently is accompanied by weakness but this is not always so. Muscle pain (myalgia) and tenderness may arise from lesions in the peripheral nerve or blood vessel rather than the muscle tissues proper. Aches and pains are frequent in many viral infections and usually have no known morphologic counterpart. Zenker's degeneration is a pattern of hyalinization of muscle originally described in febrile illnesses such as influenza and typhoid fever. Fragmentation of the myofibers and regenerative changes were associated with this. These changes are not specific and rarely are seen at the present time.

CRAMPS

Cramps, strictly speaking, are spasmodic painful involuntary contractions of muscle. They are distinguished from twitches or tics by persisting at least several seconds. Their spasmodic character excludes the sustained rigidity of corticospinal and parkinsonian types. The term also is often applied to spasmodic contractions in which pain is not a feature. The mechanism by which pain is produced is unknown. Speculations about the possible role of accumulation of lactate during contraction or of altered potassium concentrations are not supported by existing data. Cramps may arise from abnormalities of the muscle fiber or the neuron.

Ordinary muscle cramps, precipitated by exertion or by minor movements during sleep, probably arise in motor nerve fibers or spinal neurons. The pain may be regarded teleologically as a mechanism for protecting the muscle from overexertion, since it is known that excessive contraction may cause necrosis of myofibers. The EMG discloses irregular high-frequency high-voltage bursts. The tetany that complicates hypocalcemia or alkalosis reflects a lowered threshold of the motor nerve fibers. Typically the cramps of tetany are not painful. Another neurogenic form of painful cramp is the tetanus caused by the neurotoxin of *Clostridium tetani*. Initially tetanus is characterized by an increased sensitivity to stretch reflexes and ultimately the spinal neurons fire continuously.

The myogenic muscle cramps are of two sorts: muscle contracture and myotonias. Contractures cause pain and myotonias do not. Muscle contractures are contractions of muscle that persist after excitation has ceased. The EMG shows little or no activity. The contracture must therefore arise in the muscle cell distal to the postsynaptic surface of the myoneural junction, and, for this reason, there is no response to curare. Muscle contractures are to be distinguished from what are loosely called contractures following prolonged disuse or immobilization of joints: such joints become fixed in abnormal positions and passive stretching causes pain. This rigidity arises in large part from changes in the various connective tissues rather than the myofiber proper. McArdle's disease is a rare, inherited disease of muscle but is particularly interesting as a prototype because it results from a specific biochemical defect. There is a deficiency of glycogen phosphorylase in the muscle and insufficient formation of pyruvate and lactate. As a result, even mild muscular effort leads to a crampy pain, weakness, and contracture.

Myotonias are characterized by deficient relaxation of voluntary contractions of muscle. Unlike contractures, myotonic contractions are accompanied by the electrical hyperactivity patterns already noted.

INFLAMMATORY DISEASES OF MUSCLE (MYOSITIS)

When a segment of myofiber dies or degenerates, a local endomysial inflammatory response commonly occurs about the disintegrated sarcoplasm. Among the infiltrating cells are plump mononuclear phagocytes. These inflammatory foci usually are minute and scattered, but at times there are so many that differentiation between metabolic or dystrophic conditions and myositis is difficult. Minute compact aggregates of lymphocytes also are common in endomysium and perimysium, particularly about venules. These "lymphorrhages" were once considered characteristic of myasthenia gravis, but they occur in many systemic diseases and are not known to cause specific clinical dysfunctions.

Infectious Myositis

Infectious myositis most often follows introduction of bacteria through disrupted skin, *e.g.* decubitus ulcers or penetrating wounds. The local and systemic manifestations correspond to what one would expect with a particular infection. Gas gangrene is a dread complication of trauma in military and civilian practice. The principal causative agent is the Welch bacillus, an anaerobic saprophyte. It ordinarily has no effect on healthy muscle but causes extensive necrosis of macerated muscle. The proclivity of the infection to spread is related to the bacterial collagenase which dissolves the natural connective tissue barriers. Gases (mostly methane) are generated by the bacteria and the bubbles cause the tissue to be crepitant. Repair of the muscle, even if the infection is controlled, is almost non-existent.

Trichinosis is the most significant of the parasitic infestations of muscle. The disease is transmitted to man by ingestion of raw pork that contains the larvae of the nematode, *Trichinella spiralis*. Meat inspection as well as encouragement of adequate cooking have reduced the prevalence from a previous high of 22 percent of the American population 30 years ago to below 4 percent today. In only a small proportion of such persons is the infection sufficiently severe to cause symptoms, less than 200 active cases occurring each year. The figure remains much higher in some other parts of the world. Following ingestion of contaminated meat, the larvae hatch in the gut and soon cause a short-lived and mild gastroenteritis. Muscular invasion appears at the end of the first week and may continue for several weeks thereafter. Pain, tenderness, and weakness of muscles accompany the penetration of the larvae into the myofiber. There are other systemic manifestations and on occasion they may cause death. Eosinophilic leukocytosis accompanies the stage of muscular invasion. Segmental degeneration of the sarcoplasm about the larva is followed by the basophilia and nuclear changes that suggest regeneration. The endomysial sheath is preserved but intense interstitial infiltration of eosinophils and other leukocytes also is present at this time. Subsequently the larva becomes encysted within a capsule and the inflammation subsides by the third month. The cysts persist indefinitely and the capsules may slowly become calcified. Nevertheless, the larvae remain viable in the cysts for a long time. Several serologic tests have been developed but muscle biopsy remains the definitive diagnostic procedure.

Less common in western countries and usually seen in immigrants from eastern Europe are the calcified cysts of *Cysticercus cellulosae*. Cysticercus is the larval form of the pork tapeworm, *Taenia solium*. The disease is transmitted to man by the ingestion of the ova. Muscle symptoms occasionally resemble those of trichinosis but ordinarily past infection is discovered incidentally during x-ray examination for other purposes. The cysts are larger than those of trichinosis, measuring up to more than 1 cm in length and are located in the interstitial connective tissue of the muscles. Skeletal muscle fibers are parasitized with some frequency along with other organs during dissemination by the protozoa *Trypanosoma cruzi* (the cause of Chagas' disease) and *Toxoplasma gondii*. The myositis does not lead to major clinical disturbances.

Polymyositis

Polymyositis is a disease or group of diseases in which widespread degeneration or necrosis of muscle is accompanied by interstitial inflammation (Fig. 5.1). The etiology is unknown and the pathologic changes, even when severe, are non-specific. The further classification depends on the clinical context in which the polymyositis occurs. The various forms share a predilection for symmetrical involvement of the proximal muscles of the extremities. The overlying skin may be inflamed and, when skin changes are sufficiently pronounced, a separate category of dermatomyositis is distinguished. In dermatomyositis the most typical cutaneous finding is a swelling and dusky lilac-colored (heliotrope) erythema of the face and eyelids.

Polymyositis and dermatomyositis are uncommon but not rare conditions. They usually fall into the domain of the rheumatologist or neurologist. There is much variation in the clinical patterns. At one extreme is an acute form which, at its onset, has pronounced constitutional manifestations. Here fever and leukocytosis accompany the muscular symptoms. Muscle pain is less constant than weakness. These inflammatory manifestations may subside over the course of several weeks but leave a residue of permanently weak and shrunken muscle. At the other extreme is a chronic form which begins insidiously and lacks the systemic complaints. The clinical picture is dominated by slowly progressive weakness and may resemble late onset muscular dystrophies or other myopathies. These are grave diseases and carry a mortality of approximately 40 percent within a few years. The prognosis is better the younger the patient. All ages may be affected, but middle-aged adults most commonly so.

In the early inflammatory stages, necrosis and degeneration are seen in discrete segments of the myofibers rather than in entire bundles. The degeneration is hyaline or floccular and at times the sarcoplasm is vacuolated. Plump mononuclear phagocytes surround the fragmented sarcoplasm. The other infiltrating cells are predominantly lymphocytes and small mononuclear cells, there being only a small scattering of eosinophils and plasma cells. Evidences of attempted regeneration of myofibers also are seen. In more advanced stages, the inflammatory component may disappear and the residue is a markedly atrophic and fibrotic muscle. There is much variation in the diameter of the remaining myofibers. Such a fibrotic muscle is not merely weak but shortened and rigid so that there is a contracture of the joints.

In the early inflammatory phases, the various muscle enzymes are released from the damaged myofibers and appear in the serum. The most sensitive and widely used measures of muscle damage are the serum lev-

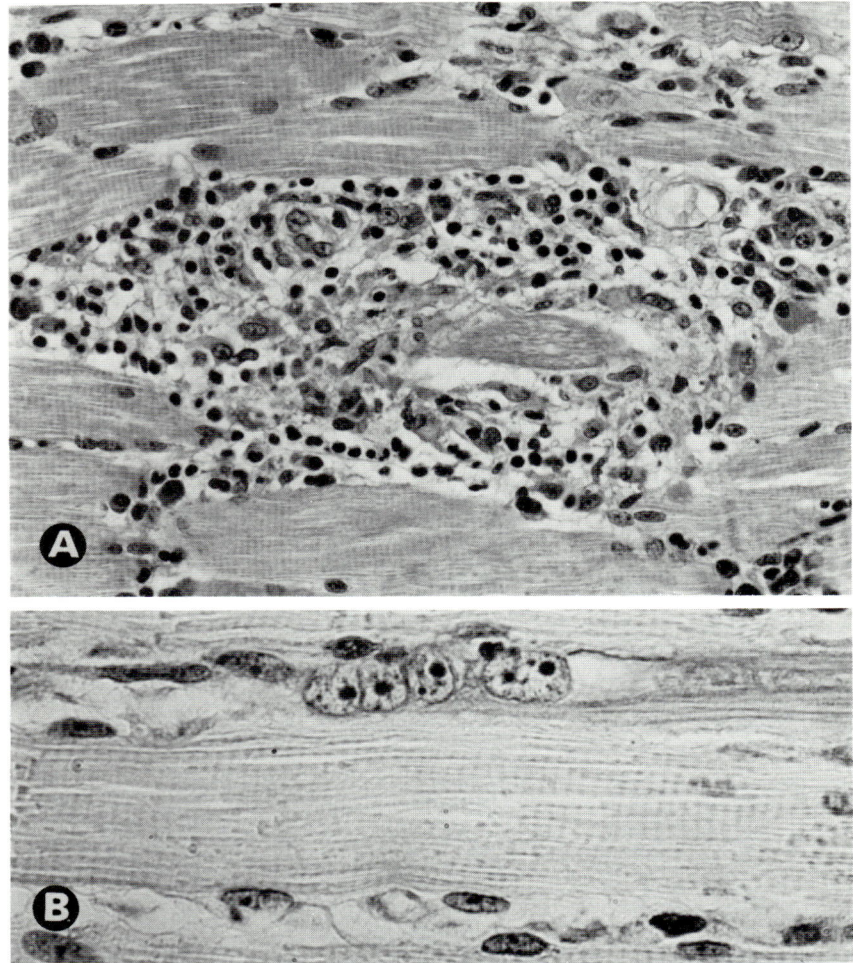

Figure 5.1. Polymyositis. *A.* An infiltrate of mononuclear phagocytes, lymphocytes, and scattered other cell types surround a necrotic muscle fiber. (Hematoxylin and eosin, × 380.) *B.* Proliferation of sarcolemmal nuclei. The nuclei are arranged in a row, are large, and have prominent nucleoli. The sarcoplasm of the fiber lacks cross-striations and has a basophilic cast indicated by the darker appearance at the *right*. (Same case, ×800.)

els of creatine phosphokinase (CPK), aldolase (from the glycolytic pathway), and the transaminases (SGOT, SGPT). These levels return to normal as the inflammatory changes burn out. The EMG changes are complex and have been recommended as a guide to selecting a site for muscle biopsy.

Approximately half the cases of polymyositis and dermatomyositis after the fifth decade are associated with malignant tumors located at a distant site. The occurrence of polymyositic syndromes in this age group always warrants search for occult carcinoma (lung, pelvic organs, and gastrointestinal tract). At times resection of the carcinoma results in remission of the muscle disease. Raynaud's phenomenon and minor arthritic manifestations occur in roughly half the patients with polymyositis, but overt rheumatoid arthritis and scleroderma are infrequent. It is tempting to speculate about autoimmune mechanisms that might be involved in the pathogenesis of polymyositis, but supporting evidence is sparse. Lesions closely resembling polymyositis are readily produced in certain experimental animals by quite different means, e.g., nutritional deficiencies or hyperadrenalcorticism.

A childhood form of dermatomyositis differs from the preceding in having prominent vascular lesions not only in the muscle but in splanchnic organs. The lumens of small arteries and arterioles are compromised by loose-textured intimal fibrous tissue or by thrombi composed of platelets and fibrin. These predispose to infection and ulceration of the gastrointestinal tract and death commonly results from perforation of the ulcers.

Myositis Ossificans

To be distinguished from true myositis are two muscle diseases in which inflammatory cell infiltration is trivial while new bone formation is prominent.

Traumatic myositis ossificans is a disorder of adolescent boys or young men. It follows injury to the muscle, the injury itself sometimes quite minor. Some weeks later, induration of the part becomes apparent and the presentation may suggest a malignant bone tumor. At its inception, the muscle is torn, usually in the vicinity of the periosteum. There follows a localized proliferation of vascular and myxoid fibrous tissue as well as hyaline cartilage and membranous bone in the endo- and perimysium. Hemorrhage is not commonly seen. It must be emphasized that, in the early stages, proliferation of mesenchymal cells is florid and too readily confused with osteogenic sarcomas. Here once again is an example of the pluripotentiality of connective tissue cells. The islands of cartilage and bone that form have an orderly, laminate disposition unlike that found in osteogenic sarcomas. The location of the lesions is quite characteristic, mostly the brachialis anticus and the quadriceps femoris. Occasionally other areas also are affected.

Progressive myositis ossificans differs in almost every way from the preceding. It is rare and a congenital rather than a traumatic disease. Extensive new bone formation takes place in the connective tissues associated with the bone—the fasciae, aponeuroses, and tendons. The parts of the body affected may thus become pathetically immobile.

MISCELLANEOUS NON-INFLAMMATORY CONDITIONS

Polymyalgia Rheumatica

Although recognized only within recent years, polymyalgia rheumatica is already diagnosed as frequently as gout in arthritis clinics. It is a disease of the elderly and the etiology is unknown. The principal symptoms are aching pain in the muscles about the shoulder or the pelvic girdle. Weakness is not a feature. Physical findings are normal as are the histologic appearances of the muscle, EMG, and serum muscle enzyme levels. The only consistent abnormality in the laboratory data is an elevated erythrocyte sedimentation rate. There is typically a prompt clinical improvement following treatment with small doses of corticosteroids.

The structural basis for the symptoms is not clear. There is a frequent association with giant cell arteritis—perhaps this is the cause of the pain. The relationship of polymyalgia rheumatica to temporal arteritis, also a giant cell lesion, is not clear. Unlike the temporal arteritis syndrome, blindness may occur but is not a common manifestation of polymyalgia rheumatica.

Myoglobinuria

When muscle fibers become necrotic, myoglobin is liberated into the body fluids. It is of relatively low molecular weight and is excreted by the kidney. The urine thereby becomes discolored a dark red-brown. Myoglobinuria is distinguished from hematuria by microscopic demonstration of the absence of red blood cells, and from hemoglobinuria and porphyrinuria by spectroscopy. As one would expect, serum levels of muscle enzymes also are elevated in myoglobinurias.

The commonest cause of myoglobinuria is the crush syndrome. This follows severe crushing injuries of muscle and is manifested principally by acute renal failure. Shock seems to be the common denominator of most forms of acute renal failure, but myoglobin casts also are regularly found in the crush syndrome and presumably contribute to its development. Renal failure also complicates other types of severe myoglobinuria. Myoglobinuria is a feature of the necrosis (rhabdomyolysis) that follows extreme muscular exertion. Paroxysmal myoglobinuria is an uncommon condition which occurs spontaneously or after only mild exertion. Such cases must be distinguished from paroxysmal hemoglobinurias. The severity of the lesion varies greatly. In some cases death results either from involvement of respiratory muscles, hyperkalemia, or renal tubular necrosis. In others, the lesion is localized to fewer muscles and is of lesser clinical consequence. The affected muscles are painful and weak. Histologically the myofibers appear hyalinized and floccular. The endomysial connective tissue sheath is preserved. If the individual survives, therefore, extensive regenerative activity is found and ultimately the muscle is completely restored.

REFERENCES

BANKER, B. Q., AND VICTOR, M. Dermatomyositis (systemic angiopathy) of childhood. Medicine 45:261, 1966.

BLAND, J. H., KIRSCHBAUM, M. B., O'CONNOR, G. T., AND WHORTON, E. Myositis ossificans progressiva. Effect of intravenously given parathyroid extract on urinary excretion of connective tissue components. Arch. Intern. Med. 132:209, 1973.

BRODY, I. A. Muscle contracture induced by exercise. A syndrome attributable to decreased relaxing factor. N. Engl. J. Med. 281:187, 1969.

CURRIE, S., SAUNDERS, M., KNOWLES, M. AND BROWN, A. E. Immunological aspects of polymyositis. Q. J. Med. 40:63, 1971.

S. E. GOULD (Ed.) Trichinosis in Man and Animals. Springfield, Ill., Charles C Thomas, 1970.

HAMILTON, C. R. Jr., SHELLEY, W. M., AND TUMULTY, P. A. Giant cell arteritis; including temporal arteritis and polymyalgia rheumatica. Medicine 50:1, 1971.

HAMRIN, B. Polymyalgia arteritica. Acta Med. Scand. Suppl. 533:1, 1972.

LAYZER, R. B., AND ROWLAND, L. P. Cramps. N. Engl. J. Med. 285:31, 1971.

PEARSON, C. M. Polymyositis and dermatomyositis. *In* Immunological Diseases, 2nd ed., Vol. 2, p. 1039, edited by M. Samter. Boston, Little Brown, 1971.

SAVAGE, D. C. L., FORBES, M., AND PEARCE, G. W. Idiopathic rhabdomyolysis. Arch. Dis. Child. 46:594, 1971.

WEINSTEIN, L., AND BARZA, M. A. Current concepts; gas gangrene. N. Engl. J. Med. 289:1129, 1973.

WHITAKER, J. N., AND ENGEL, W. K. Vascular deposits of immunoglobulin and complement in idiopathic inflammatory myopathy. N. Engl. J. Med. 286:333, 1972.

6

THE WEAK MUSCLE

Some of the difficulties in distinguishing true myopathies from motor neuropathies have already been indicated. These difficulties have been compounded by recent histochemical evidences of neurotrophic influences on muscle properties in disease. A muscle may be regarded as weak not only when it fails to contract adequately but also if it is myotonic and fails to relax. Fatigue is not a single but a multifaceted phenomenon in which a muscle progressively loses its ability to contract after exertion. In addition, there is a subjective sensation of fatigue or weariness. Alterations in the electromyography (EMG) patterns occur during fatigue but they are not distinctive and may reflect circulatory changes in the muscle.

ATROPHY

The turnover of muscle protein is very rapid, the half-life being about 10 days. The rate is influenced by nutritional and hormonal factors but also depends on the work the muscle has to carry out. The control mechanism by which the mechanical stimulus is translated into the biochemical events is not clear. Skeletal muscle has a remarkable capacity for work hypertrophy. The type of work makes a great deal of difference, intensive exercise more likely to cause bulging muscles than sustained effort. The obverse, atrophy of muscle, is effected first by a reduction in the number of myofilaments within the myofiber. Failure to use a muscle, such as follows surgical splinting, results in grossly visible atrophy within a few days. Atrophy and degeneration of muscle are two separate phenomena. Isolated degenerated fibers are sometimes seen in atrophic muscle, particularly in advanced stages, but they are secondary abnormalities. They probably result from abnormal contractions imposed by the atrophy. In longstanding and severe atrophy, many muscle fibers disappear. Uncomplicated atrophy does not lead to fibrosis although an apparent histologic increase of endomysial and perimysial collagen may result from condensation of the stroma as the muscle disappears. The presence of extensive fibrous tissue in a biopsy section suggests an antecedent inflammatory or other sclerotic process. On the other hand, extensive replacement of muscle by adipose tissue is common in protracted atrophy. Only the presence of a few myofibers may betray the original nature of the tissue. Adipose replacement is most pronounced in the Duchenne type of muscular dystrophy but it also is seen in muscles paralyzed by neurologic disease or by disuse.

Atrophy is manifested histologically by a reduction in the diameter of the myofibers and an apparent increase in the number of sarcolemmal nuclei. The increase actually results from the loss of muscle protein substance, while the nuclei persist. These changes are seen more easily in transverse than longitudinal sections. When the tissue specimen has been fixed isometrically, the atrophic fibers appear more angular than normal. The cross-striations are preserved. Atrophy of muscle bundles, as distinct from the individual myofibers, may follow different patterns, depending on the underlying cause.

MOTOR NEUROPATHIES

Distal motor syndromes, those in which weakness and atrophy affect principally the muscles of the feet and legs or of the hands and forearms, usually result from disease of the lower motor neuron. They may arise in the peripheral nerve or centrally in the spinal cord (*e.g.*, anterior poliomyelitis). The characteristic pattern of non-specific longstanding lower motor neuron atrophy is for all or almost all of the fibers within a fascicle to be reduced to small size. This pattern is different from proximal muscle syndromes, those affecting the shoulder or pelvic girdles. The latter syndromes are more usually polymyositis or certain types of muscular dystrophy.

In the earliest stages of neuropathic muscle atrophy, the above pattern is not seen. Often the process of atrophy within the fascicle is not uniform but small fibers are scattered among those of normal size. Histochemical studies disclose that the atrophy has occurred preferentially in the Type II fibers. A similar pattern is found in disuse and several other sorts of atrophy (*e.g.*, cachexia and corticosteroid-induced). Heavy weight is placed on these findings by Engel in believing that the latter varieties of atrophy are in fact mediated trophically through the nerves rather than being a direct response of the myofibers. Preferential atrophy of Type I fibers is less common and is fairly characteristic of myotonic dystrophy. It is a generally useful clinical rule that serum levels of muscle enzyme activities are not elevated in the motor neuropathies.

There are many types of peripheral neuropathy and they may or may not involve sensory as well as motor fibers. The lesions may affect either the axon or the Schwann cell sheath. Segmental demyelination is pathognomonic of Schwannian neuropathies. It causes greater slowing of motor nerve conduction than does axonal degeneration. Most of the common causes of peripheral neuropathy (alcoholism, thiamine deficiency, poisoning by lead, organic mercury, and triorthocresylphosphate) are of axonal type. The delayed polyneuropathy that sometimes follows diphtheria, on the other hand, is due to demyelination caused by the diphtheria toxin. The Guillain-Barré syndrome is a well known, although uncommon acute idiopathic polyneuropathy associated with a high concentration of protein but no excess of cells in the spinal fluid. It often follows a viral infection or inoculation procedure and may therefore be an analogue of post-infectious allergic demyelinating encephalopathy. This condition may last for several months and then disappear as a new myelin sheath is generated. Sometimes death results from paralysis of respiratory muscles.

MUSCULAR DYSTROPHIES

These are a heterogeneous group of genetically governed progressive degenerative diseases of voluntary muscle. There are many varieties, most of them infrequent. The nature of muscular dystrophies is obscure and accordingly it is not known whether they are related to each other pathogenetically. For this reason they are distinguished principally by clinical features such as the age of onset, the location of the muscles affected, and the mode of inheritance. It is estimated that there are 200,000 cases of muscular dystrophy in the United States. For none is there a satisfactory treatment.

The *pseudohypertrophic muscular dystrophy of Duchenne* is by far the most frequent. The symptoms usually appear insidiously during the first years of life. The child, previously normal, begins to walk clumsily, falls down, and has difficulty getting up without assistance. The weakness progresses relentlessly and comes to involve the trunk and upper extremities. Ultimately the patient becomes bedridden, contractures develop, and death usually results by the second decade. There are no abnormalities of nerve conduction. The condition is usually inherited as a sex-linked recessive and therefore occurs only in boys.

The muscle in the Duchenne dystrophy, despite the pathetic weakness, clinically appears enlarged and firm rather than withered (Fig. 6.1). This paradox results from the marked adipose replacement of the atrophic muscle. In its early phases, the histologic changes are those of non-specific degeneration. Individual fibers are necrotic. Muscle cells, not yet affected, generally have a normal appearance and complement of enzymes. Low grade ultrastructural abnormalities may, however, be found in the cells and in those of asymptomatic carriers of the trait. Basophilic fibers with enlarged sarcolemmal nuclei are seen in small numbers

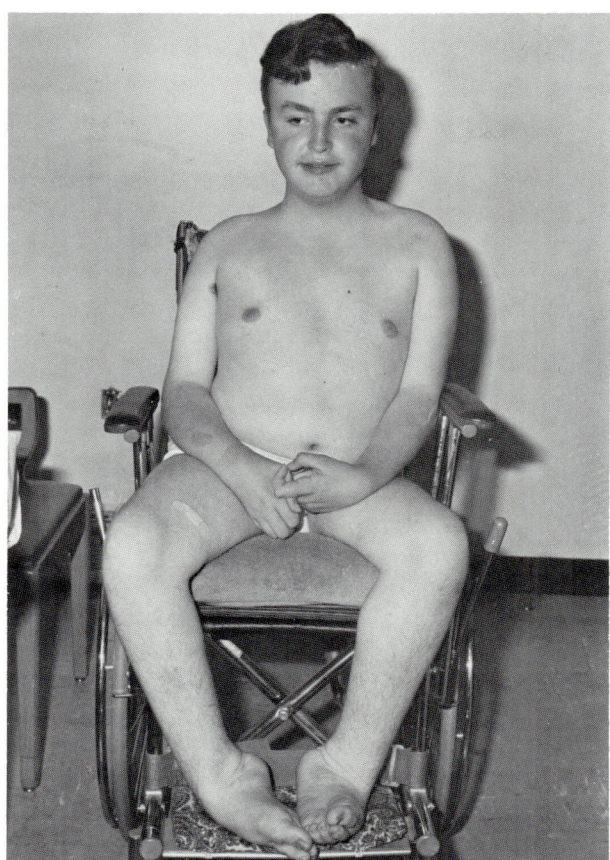

Figure 6.1. Duchenne dystrophy in a 14-year-old boy. Although muscle weakness has confined him to a wheel chair and caused talipes-like contracture of the feet, the muscles appear enlarged.

and the concept that the basic defect in this dystrophy is a lack of regenerative capability is weakened thereby. Atrophy and adipose infiltration follow with a variable degree of endomysial condensation fibrosis.

The serum levels of creatine phosphokinase, aldolase, and transaminases are elevated in proportion to the rapidity of muscle destruction. In the advanced stages, these levels fall off. Creatine phosphokinase (CPK) levels also are somewhat elevated in unaffected carriers of the trait. This feature may distinguish spontaneous mutations from the inherited form and thus be of great value in genetic counseling. Another measure of muscle destruction is the alteration in creatine metabolism. Because the muscle fibers no longer are available to catabolize the creatine that the liver synthesizes, the amount of creatine excreted in the urine increases greatly while the creatinine levels are greatly reduced. These metabolic disturbances necessarily increase as the muscle damage progresses.

There are several variants of the Duchenne type of muscular dystrophy, perhaps as many as one-third departing from the preceding picture. In some youngsters, the age of onset is later (6 to 18 years). The course of the illness is more benign and the life span is almost normal. There also exist autosomal (*i.e.*, genes not located on the sex chromosome) types that can affect girls as well as boys.

Other types of muscular dystrophy are less common and generally less incapacitating. Brief mention is made of only two of these and the interested reader is referred to Walton for further information on the subject. *The facioscapulohumeral dystrophy of Landouzy and Dejerine* is inherited as an autosomal dominant. It may appear at any age, although generally during the 2nd to 4th decade, and it affects principally the proximal muscle groups of the arms as well as the facial muscles. *Myotonic dystrophy* is a more complex disorder where, in addition to the weakening and wasting, difficulty in relaxation of muscle also is a problem. One histologic feature of this as well as other myotonic diseases is the occurrence of ring fibers, *i.e.*, myofibers in which a proportion of myofibrils course spirally instead of axially. Some cross-striations are thus seen as concentric rings in transverse sections. Another curious feature of myotonic dystrophy is a predilection for Type I fiber atrophy.

MYASTHENIA GRAVIS

Myasthenia gravis is a chronic disease characterized by excessive weakness of voluntary muscle following use. The symptoms are therefore most troublesome toward the end of the day. Although the paralysis follows exertion, it is not accompanied by a sense of fatigue. The muscles of the head (ocular, facial, swallowing, and speech) are the ones principally affected but the neck, trunk, and limbs may also be involved. Myasthenia gravis is principally a disease of adult life and follows a fluctuating course. Long remissions may occur. Although the muscles are weak, they do not undergo the severe wasting found in the preceding motor neuropathies or myopathies. The EMG changes are of a "myopathic" type: short duration, small amplitude, and high frequency potentials.

These manifestations result from a deficiency of acetylcholine at the myoneural junction. The definitive diagnostic feature of myasthenia is a transient improvement following administration of cholinesterase inhibitors such as prostigmine analogues. The myasthenia *a priori* might result from a defect in the presynaptic release of acetylcholine or a competitive block in the postsynaptic receptors. The affected muscles in time become somewhat atrophic but otherwise are normal. Scattered lymphorrhages (see Fig. 4.4) are present in the endomysium but are not specific for the disease. Structural abnormalities of the motor end plate are found with the electron microscope, principally a loss of postsynaptic folds. Excessive branching of the terminal axonal filaments also exists at times. Both these changes are also seen at certain stages following denervation and therefore do not necessarily indicate a fixed structural defect in myasthenia gravis.

In a large proportion of cases of myasthenia gravis, there are thymic abnormalities. Approximately 30 percent of the patients have a thymoma, principally of spindle cell or epithelioid type. A variable degree of hyperplasia is present in the majority of the others while, in the remainder, lymphoid follicles bear witness to an abnormality of the thymus. Several types of circulating antibodies occur in this condition, although not always the same ones in all cases. Some are directed against muscle proteins, others against thymic epithelial cells; in a smaller proportion there are anti-DNA or antithyroid antibodies in the serum. They may thus be considered autoantibodies and this situation is reminiscent of rheumatic diseases such as systemic lupus erythematosus and Sjögren's syndrome.

These antibodies presumably are synthesized in the various lymphoid organs, including the thymus and lymphorrhages. It is conceivable that the antibodies mediate the synaptic insufficiency. These considerations underlie proposed therapies with corticosteroid or immunosuppressive agents as well as the parasympathomimetic compounds. Thymectomy is beneficial in some but not all cases. At one time myasthenia gravis caused death in 85 percent of cases within 2 years. This figure has been reduced greatly by the anticholinesterases.

MISCELLANEOUS MYOPATHIES

Numerous other muscle diseases can be recognized by distinctive clinical, morphologic, or biochemical abnormalities. They are quite rare and have been reviewed by Engel.

Episodic diseases include *paroxysmal familial hypokalemic paralysis*. In this disorder, periods of paralysis or motor weakness lasting several hours result from a shift of potassium from the extracellular fluid into the myofiber. The serum potassium levels may decline to as low as 2 meq per liter. The sarcoplasm becomes coarsely vacuolated. The underlying defect must be a fault in the permeability of the sarcolemmal membrane.

The *floppy baby syndrome* is a generic term for many unrelated types of congenital muscular weakness in infants. Their etiology varies from absence of anterior horn cells (infantile spinal muscular atrophy or Werdnig-Hoffmann disease) to a series of myopathies that have distinctive biochemical or morphologic abnormalities. Several types of *glycogen storage disease* of muscle may result from specific enzyme deficiencies.

McArdle's disease, already mentioned, is a relatively benign form of glycogenosis (Type V) related to absence of glycogen phosphorylase. In Pompe's (Type II) disease, the defect is principally in the lysosomal enzyme, acid maltase. Death occurs in early infancy. In von Gierke's disease (Type I), muscle is spared because this tissue normally does not contain the enzyme involved, glucose 6-phosphatase. Rod and central core myopathies are so designated because of their histologic appearance.

Weakness and wasting of proximal muscles are well recognized in metabolic diseases. They may even dominate the picture of thyrotoxicosis and disappear when the hormonal disturbance is corrected. There are no distinctive histologic changes in the muscle here. Muscle weakness also is a feature of Cushing's syndrome and may be an untoward consequence of corticosteroid therapy in other diseases. Muscles also commonly atrophy in arthritis because of disuse but in many cases of rheumatic disease weakness is so severe as to require some other explanation. Occasionally more severe changes associated with myofiber damage and inflammatory infiltration resemble polymyositis; this is more often seen in scleroderma and systemic lupus erythematosus than in rheumatoid arthritis.

REFERENCES

BALOH, R., CANCILLA, P. A., KALYANARAMAN, K., MUNSAT, T., PEARSON, C. M., AND RICH, R. Regeneration of human muscle. A morphologic and histochemical study of normal and dystrophic muscle after injury. Lab. Invest. 26:319, 1972

ENGEL, W. K. (Ed.) Current concepts of myopathies. Clin. Orthop. 39:2, 1965.

FISHER, E. R., WISSINGER, H. A., GERNETH, J. A., AND DANOWSKI, T. S. Ultrastructural changes in skeletal muscle of muscular dystrophy carriers. Arch. Pathol. 94:456, 1972.

McFARLIN, D. E. Myasthenia gravis. *In* Immunological Diseases, 2nd ed., Vol. 2, p. 1150, edited by M. Samter. Boston, Little Brown, 1971.

SCHOTLAND, D. L. Ultrastructure of muscle in glycogen storage diseases. *In* The Striated Muscle, p. 410, edited by C. M. Pearson and F. K. Mostofi. Baltimore, Williams & Wilkins, 1973.

7

DISTURBED SKELETAL GROWTH

Abnormalities of growth are frequent and may affect the entire skeleton or be confined to parts of it. A child's limb fails to grow, in the simplest instance, because an injury causes the epiphyseal plate to become vascularized and close prematurely. A comparable process may complicate arthritic diseases in children. Involvement of the temporomandibular joint in juvenile rheumatoid arthritis, for example, limits growth of the mandible from the condylar cartilage and the jaw thereby becomes underslung (micrognathia). When the growth plates near many joints are so affected, the limbs may be stunted, a phenomenon called rheumatic dwarfism. A different mechanism restricting growth in one locality is neural deprivation. The commonest forms of this, both fortunately now less frequent than just 3 decades ago, are anterior poliomyelitis and Erb's palsy. Erb's palsy results from forceful overstretching of the head and shoulder of an infant during birth. The fifth cervical nerve is damaged thereby and the forearm, in addition to being partly paralyzed, is shortened. The explanation usually offered for this is that trophic effects, normally exerted by nerves on non-muscular tissues of the limb, are missing.

In other common types of localized growth disturbance leading to deformity it is not clear that either of the above mechanisms operate. For example, idiopathic scoliosis (lateral curvature of the spine), with or without a humpback (kyphosis), usually makes its appearance in early adolescence. Weakness of some of the muscles that normally maintain balance of the growing vertebral column is postulated to be a factor but a neural explanation, such as isolated poliomyelitis, is rarely found. In accordance with Wolff's law, the abnormal distribution of pressure causes portions of the vertebral bodies in the affected region to fail to grow.

Most types of growth disturbance, however, are not of local origin. Skeletal growth is interrupted during periods of debility such as illness or malnutrition. When this happens, the basal columns of the growth cartilage become shortened and a transverse layer of bone is deposited just beneath them. The epiphysis is thereby sealed off from the metaphysis but ossific union of the secondary center and shaft of the

74   *Musculoskeletal System*

bone does not occur. When the debility is relieved, growth can therefore resume and the epiphysis moves away from this abnormal layer of bone. The latter persists for a considerable period of time and appears in x-ray films as a transverse growth arrest line (Fig. 7.1).

SHORT STATURE

More than a million American children are abnormally short. This causes serious emotional distress and as a result is the commonest presenting complaint in pediatric endocrine clinics. Many youngsters are

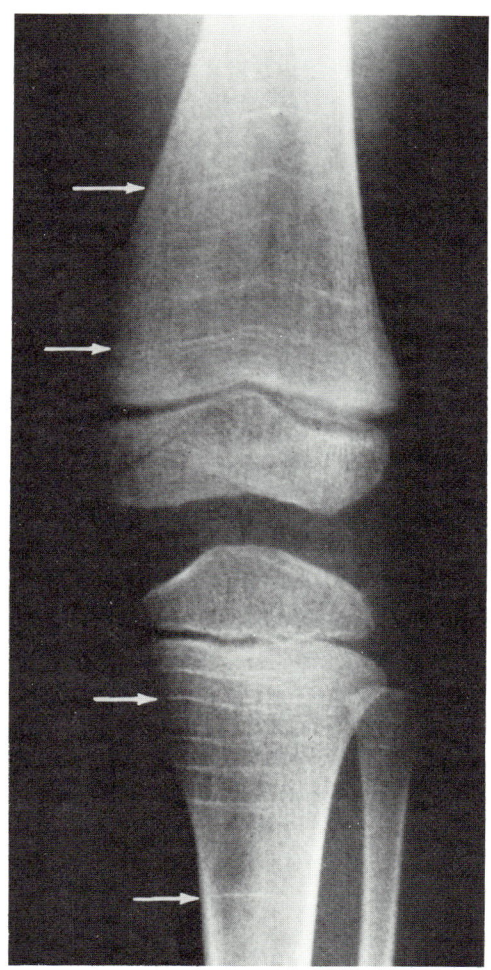

Figure 7.1. Growth arrest lines in a 14-year-old girl with episodic chronic illness.

late bloomers and outgrow the problem. Others inherit their shortness and do not. Much of what was once considered genetic small stature ultimately has proven not to be. The mean height of native born American white males increased roughly an inch per generation during the 19th century, perhaps because nutrition and general health improved. How short is short? Eligibility for membership in the Little People of America is an adult height of not more than 4 feet and 2 inches. The Association has 20,000 members. A small stature may affect all parts of the skeleton proportionately or only certain parts and therefore be disproportionate. Midgets are persons in the former category. The term dwarfism is more often applied to disproportionate short stature.

In the clinical analysis of shortness, chronologic, height, and bone ages are compared (Fig. 7.2). The latter two values are the ages of a normal population having, respectively, height and epiphyseal maturation comparable to that of the patient. The wrist and the hand are the site usually examined radiologically. Standard values for osseous development presented in the atlas of Greulich and Pyle have stood the test of time.

More than 250 varieties of short stature are known and they have been classified many ways. Some arise within the skeleton or its primordium and thus are chondrodystrophies or osteochondrodystrophies. In others, the fault lies in an extraskeletal regulatory mechanism.

Among the intrinsic varieties of short stature, the most frequent is *achondroplastic dwarfism*. Here the limbs are short and the frontal part of the head is excessively large. The trait is inherited as an autosomal dominant. Although relatively little histologic or chemical abnormality has been found in the epiphyseal cartilage of achondroplastic dwarfs, its rate of growth and endochondral ossification is reduced. *Multiple epiphyseal dysplasias* (and if the spine also is involved, spondyloepiphyseal dysplasias) are another group of relatively frequent inherited chondrodystrophies. Stature is only moderately reduced and much more proportionate than in the preceding disorder. There are several genetic patterns. These disorders predispose to degenerative joint disease and severe osteoarthritis of the hip occurs at an early age generation after generation.

Well characterized metabolic disturbances exist in the six major varieties of *mucopolysaccharidoses* (p. 8). The chemical abnormalities are identified by the appearance of excessive quantities of one or another glycosaminoglycan in the urine. The cells of the cartilaginous skeleton are involved along with those of other connective tissues in the storage of these materials, and dwarfism is a feature of all except the rare Schei's and Sanfilippo's syndromes. In the Morquio-Brailsford mucopolysaccharidosis, unlike the others, skeletal manifestations dominate the

76 *Musculoskeletal System*

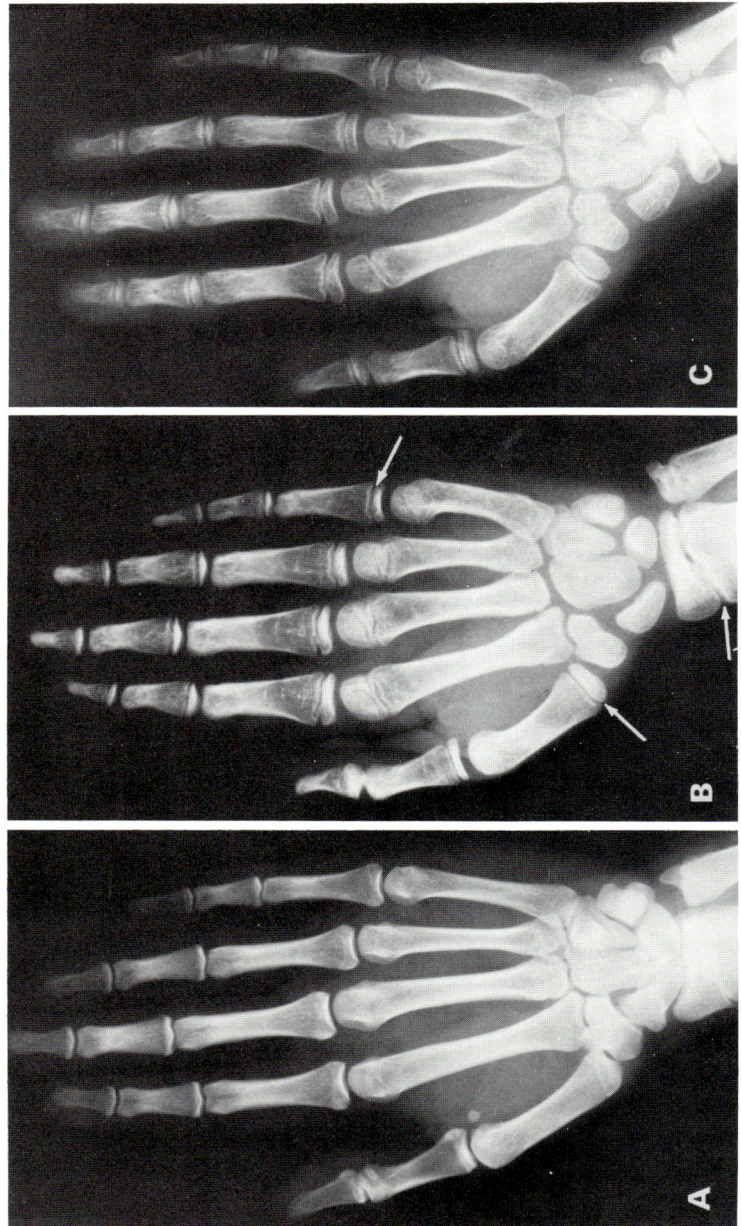

Figure 7.2. Retarded bone age in a 15-year-old hypothyroid child (*B*). Although the chronologic age is the same as that of a normal 15 year old (*A*), the bone age indicated by the shadows of the ununited epiphyses (*arrows*) is comparable to that of a normal 9-year-old child (*C*).

clinical picture. The bones are misshapen in addition to being short. Flattening of the vertebral bodies is a major radiologic finding. Persons surviving to adulthood are prone to degenerative joint disease. The urine contains excessive quantities of keratan sulfate. The disorder is transmitted as an autosomal recessive. In Hurler's disease, the defective enzyme is L-iduronidase. Connective tissue cells store and secrete excessive amounts of dermatan sulfate that they cannot degrade. Abnormal quantities and types of connective tissue matrices are formed in the skeleton and cardiovascular system, and there is marked stunting of growth. Mental retardation with progressive hydrocephalus is the principal manifestation of Hurler's disease. It is not clear how the cerebral lesions are related to the connective tissue defect.

In other sorts of chondrodystrophoid dwarfism, there are structural abnormalities of cartilage but as yet no known biochemical defects. Specific disturbances in the oxidation of glucose or the synthesis of chondroitin sulfate by chondrocytes have been found in chondrodystrophoid diseases of other species. One awaits developments of this sort in the human disorders.

Endocrinopathies

Disturbances of all the endocrine glands except the adrenal medulla may result in short stature. The most important of these is anterior hypopituitarism. This usually is a sporadic rather than an inherited condition. The most frequent cause of acquired hypopituitarism in childhood is a craniopharyngioma. This is a tumor that arises in Rathke pouch remnants at the base of the brain, and compresses the anterior lobe of the pituitary. There are two principal patterns depending on whether the deficiency is confined to growth hormone (GH) or also involves others such as gonadotropins or thyrotropin. In isolated GH deficiencies, shortness is of proportionate type. The height age is reduced but not the bone age. Such children can respond to the administration of human GH prior to but not after the age of sexual maturity. Serum levels of GH and somatomedin are low. When gonadotropins also are deficient, sexual development is retarded. Accordingly bone age remains low, epiphyses often being ununited well into the 4th decade. In such instances, the potential for response to human GH persists for a longer period of time than in the isolated GH deficiencies. African pygmy tribes have normal serum levels of GH. One must assume that their short stature results from some non-responsiveness to this hormone. The growth hormones of various animal species differ chemically. It is not possible to treat hypopituitary dwarfism with GH of other species because the latter are physiologically inactive in man and because they produce allergic side reactions. Human GH necessarily is in short supply. It is possible that somatomedins, which are much smaller molecules, will find their way into the therapy of hypopituitarism.

Androgens have two distinct and somewhat contradictory effects on the skeleton. They are general anabolic agents and therefore stimulate growth of bone as well as other tissues. At the same time, they promote closure of epiphyses and the net result may be to stop growth prematurely. In sexual precocity, whether it arises from tumors secreting excessive quantities of sex hormones (chorioepitheliomas, Leydig cell or adrenal virilizing tumors), shortness and advanced bone age are a characteristic finding. Although this is a rare phenomenon spontaneously, it should be kept in mind because it may also be a by-product of treatment with sex hormones. Methyltestosterone and other synthetic androgens are frequently employed in the management of short stature in children, particularly because human GH is difficult to come by. These compounds carry a distinct risk, if used in high doses and for a long time, of augmenting epiphyseal closure and thereby exaggerating the shortness irrevocably. Sexual prematurity of girls having ovarian granulosa cell tumors also leads to short stature.

Uncomplicated hypothyroidism in children, whether cretinous or acquired, greatly retards growth and epiphyseal maturation of bone. In the acquired type, secondary centers appear but have abnormal patterns of ossification that are of diagnostic value in the x-ray films ("epiphyseal dysgenesis"). These areas of radiopacity in the epiphyses are patchy and irregular. Many manifestations of hypothyroidism respond to treatment with thyroxin, particularly if it is initiated early. Growth may at first increase dramatically but ultimately it is outstripped by epiphyseal closure. As a result, the treated child rarely attains a normal height.

Corticosteriod therapy in childhood also frequently reduces growth. As in so many other hormonal disturbances, it is difficult to know how much of this is a direct effect and how much is secondary to altered feedback to another endocrine organ. There is evidence that corticosteroids diminish GH secretion, and this may be the mechanism involved. Although insulin has a general growth-promoting action on developing tissues, growth usually is normal in well controlled juvenile diabetes mellitus.

INCREASED STATURE

Endocrine disturbances may also lead to increased height. The well known large size of eunuchs, for example, comes about because there is, in the absence of androgens, a great delay in epiphyseal closure. This permits protracted growth of the extremities.

The skeletal consequences of excessive quantities of GH depend largely on the skeletal age of the individual. Eosinophilic adenomas of the pituitary gland rarely appear before the end of the 1st decade. If they arise before puberty, they give rise to gigantism. The growth curve becomes steep and continues so until the epiphyses unite. In some cases, the tumor also interferes with sexual development and a eunuchoid component to limb enlargement is added. In itself, GH does not contribute to epiphyseal closure, and secondary centers of ossification may remain ununited in such individuals for several decades. When the pituitary adenoma appears after the epiphyses have united, growth necessarily is confined to the articular cartilages and to periosteal membranous new bone formation. Because there are so many joints in the hands and feet compared with the more proximal parts of the limbs, the increase in the length of the bones is most marked in the digits; hence the acromegaly (literally distal extremity enlargement). The mandible juts forward (prognathism) in part because it grows from the condylar cartilage, but also because, as elsewhere in the face, there is much new periosteal bone formation. Hypertrophy of soft connective tissues makes its own contribution to enlargement of the extremities and face in acromegaly. Many adenomas ultimately "burn out," *i.e.*, become hormonally inactive, or undergo cystic degeneration. If this happens at the time of puberty, gigantism is not complicated by acromegaly. If it does not, an acromegalic component is superimposed. Radioimmunoassay is a reliable diagnostic measure of GH levels in the plasma. It is available in many laboratories. Somatomedin assays, which conceivably may be more relevant, are not. Although plasma levels of GH are elevated in acromegaly, there is no linear relation between them and the clinical severity of the acromegaly. Most eosinophilic adenomas continue to be active for many years and are difficult to treat. Endocrine medications have not as yet been found generally satisfactory in lowering the GH levels. Surgical extirpation and cryohypophysectomy are the surest ways of removing the tumor, but often are incomplete and fail to control the disease. Removal of the tumor is, of course, indicated when it threatens vision. Some benefit is reported, too, from radiation therapy.

Marfan's syndrome is an uncommon inherited disorder of connective tissues affecting principally the skeletal and cardiovascular systems. It also is associated with dislocation of the lens of the eye. The principal skeletal abnormalities are excessive length of the extremities and kyphoscoliosis. As a result there is a disproportionately low ratio of the upper to the lower segments of the body, *i.e.*, the portions above and below the top of the pubic symphysis. Aneurysms of the aorta, diffuse or dissecting, and valvular insufficiency are common. This suggests that elastic as well as collagenous fibers are defective, but distinct chemical abnormalities in Marfan's syndrome have not been identified. The condition is compatible with a long life. It is transmitted as an autosomal dominant.

REFERENCES

BAILEY, J. A., Disproportionate Short Stature. Diagnosis and Management. Philadelphia, W. B. Saunders, 1973.

BLUESTONE, R., BYWATERS, E. G. L., HARTOG, M., HOLT, P. J. L., AND HYDE, S. Acromegalic arthropathy. Ann. Rheum. Dis. 30:243, 1971.

DORST, J. P., SCOTT, C. I., JR., AND HALL, J. G. The radiologic assessment of short stature-dwarfism. Radiol. Clin. North Am. 10:393, 1972.

GOTLIN, R. W., AND MACE, J. W. Diagnosis and management of short stature in childhood and adolescence. Curr. Probl. Pediatr. 2 (Nos. 4 and 5), 1972.

PYLE, S. I., WATERHOUSE, A. M., AND GREULICH, W. W. A Radiographic Standard of Reference for the Growing Hand and Wrist. Chicago, Press of Case Western Reserve University, 1971.

RIMOIN, D. L., MERIMEE, T. J., RABINOWITZ, D., CAVALLI-SFORZA, L. L., AND MCKUSICK, V. A. Peripheral subresponsiveness to human growth hormone in the African pygmies. N. Engl. J. Med. 281:1383, 1969.

SOYKA, L. F., BODE, H. H., CRAWFORD, J. D., AND FLYNN, F. L. JR. Effectiveness of long-term human growth hormone therapy for short stature in children with growth hormone deficiency. J. Clin. Endocrinol. Metab. 30:1, 1970.

VAN WYK, J. J., UNDERWOOD, L. E., LISTER, R. C. AND MARSHALL, R. N. The somatomedins. A new class of growth-regulating hormones? Am. J. Dis. Child. 126:705, 1973.

8

OSTEOPENIAS

Osteopenias are disorders in which the skeleton lacks sufficient mineral to meet its mechanical requirements and therefore is prone to fracture or deformity. The mineral phase alone or both the mineral and organic components of the matrix may be diminished. The bone may not have formed properly in the first place, become destroyed by inflammatory or neoplastic processes, or be deranged metabolically. Three principal patterns of metabolic bone disease are known: osteoporosis, osteomalacia, and osteitis fibrosa (Fig. 8.1). Osteopenias frequently are accompanied by major disturbances in electrolyte metabolism. Extensive demineralization of the skeleton may exist and still escape detection in conventional x-ray films. It is estimated that 40% of the bone's calcium must be depleted before osteoporosis can be recognized by this means.

OSTEOGENESIS IMPERFECTA

This is one of the commonest of the inherited defects of connective tissue. It is characterized by hypoplasia of bone and, to a lesser extent, of other collagenous tissues. There are two varieties: a congenital form manifest in the newborn, and a tardive form where fractures first appear later in childhood (Fig. 8.2). Death often occurs early in the former and the prognosis is better in the latter. The scleral coat of the eye also is unusually thin and, because the choroidal pigment beneath it shows through, appears blue. Hence osteogenesis imperfecta also has been called blue sclerae and brittle bone disease.

Not only the amount but the type of bone matrix formed is abnormal. Osteoblasts in the cortex have several fine structural abnormalities, but the number and electron microscopic appearance of the osteocytes are normal. The cortex is thin. Much of the bone that is formed is woven rather than lamellar. Fractures often are provoked by only slight trauma and may cause surprisingly little pain or shock. They do, of course, cripple the patient. As a result of the fractures, there may also develop grotesque deformities and shortening of the extremities, and the head

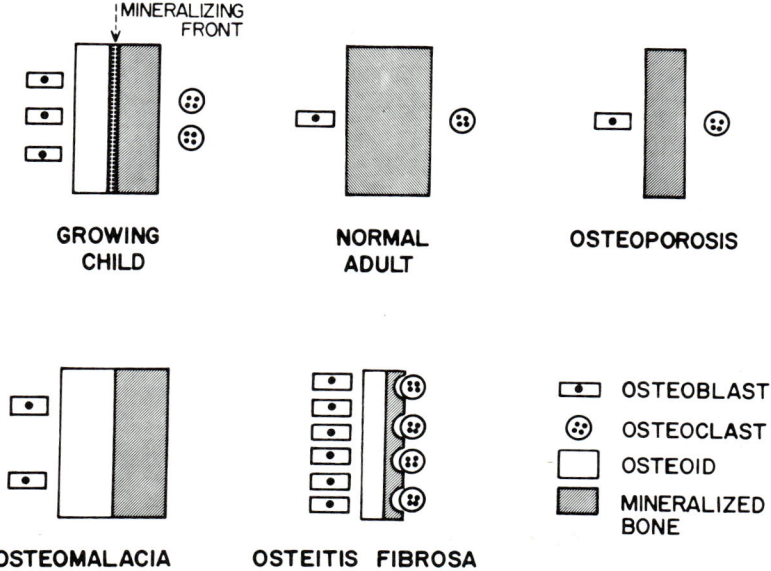

Figure 8.1. Major patterns of osteopenia, modified from Albright.

may also become misshapen. It is thus understandable that osteogenesis imperfecta is often confused with achondroplasia. The fractures heal and sometimes even form excessive quantities of bony callus. The likelihood of fractures decreases after puberty. The dentin also is hypoplastic; the teeth therefore are abnormally colored (amber or blue-gray), and prone to break or become carious. Other than elevated serum alkaline phosphatase levels during healing of fractures, no distinctive chemical abnormalities have been recognized in body fluids or bone.

Osteogenesis imperfecta is usually inherited as an autosomal dominant. Spontaneous cases sometimes arise as mutants, and recessive forms are rare. There is no generally effective systemic treatment. Management involves orthopedic measures and genetic counselling.

OSTEOPOROSIS

Although the term is frequently used loosely in roentgenologic interpretation, osteoporosis is equivalent to atrophy of osseous tissue. The name indicates that the bone is more porous than normal; that is, the volume of bone as organ remains constant but the quantity of bony tissue within it is reduced. The organic and inorganic phases diminish in equal proportion (Fig. 8.1). There are several types of osteoporosis, some of which are exceedingly frequent and important clinically.

In an extremity deprived of mechanical stimulation, not only the

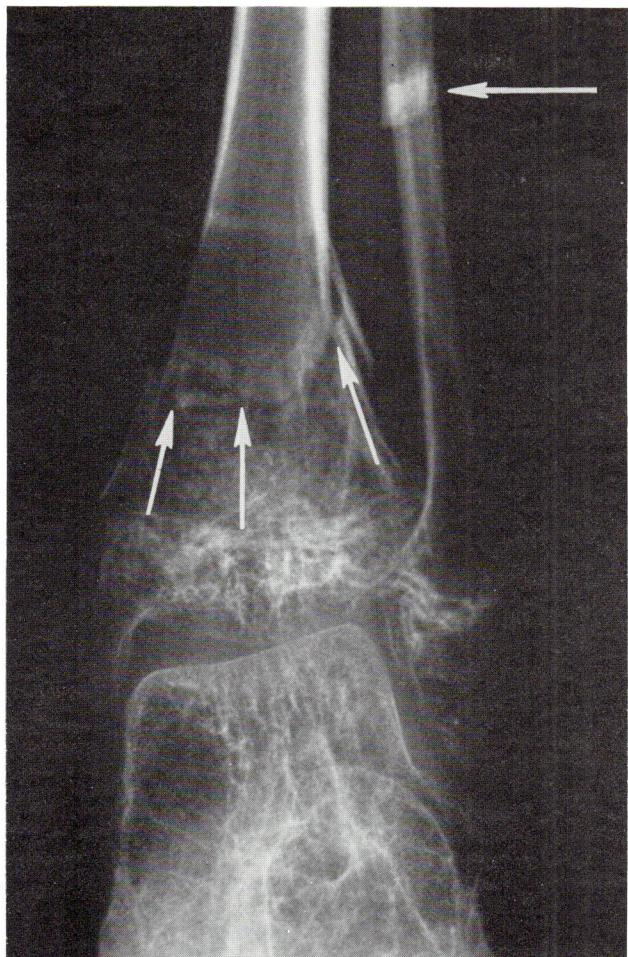

Figure 8.2. Osteogenesis imperfecta, tardive type. The cortices of the bones are very thin and multiple fractures are seen in the distal tibia and fibula (*arrows*).

muscle but also the bone atrophies thus complying with Wolff's law. Disuse atrophy commonly occurs in localized areas, as when a limb is immobilized following injury, paralysis, or arthritis. Simply putting a healthy young person to bed causes a degree of generalized osteoporosis; up to 2 percent of the body calcium is lost in 7 weeks. Disuse osteoporosis similarly is a hazard for astronauts exposed to weightlessness for protracted periods and vigorous exercises are carried out to prevent this.

Another localized post-traumatic type of osteoporosis (Sudeck's atrophy) results not so much from immobilization as from a disturbed sympathetic neurovascular reflex (Fig. 8.3). Sensory and vasomotor abnormalities occur in the distribution of the nerve. There may be, in addition to the osteoporosis, a great deal of pain and swelling in the vicinity of the joints. This may simulate arthritis. A somewhat similar picture may arise without antecedent injury. The shoulder-hand syndrome, for example, is another reflex dystrophy and it frequently follows myocardial infarction rather than trauma. Not only are there soft tissue

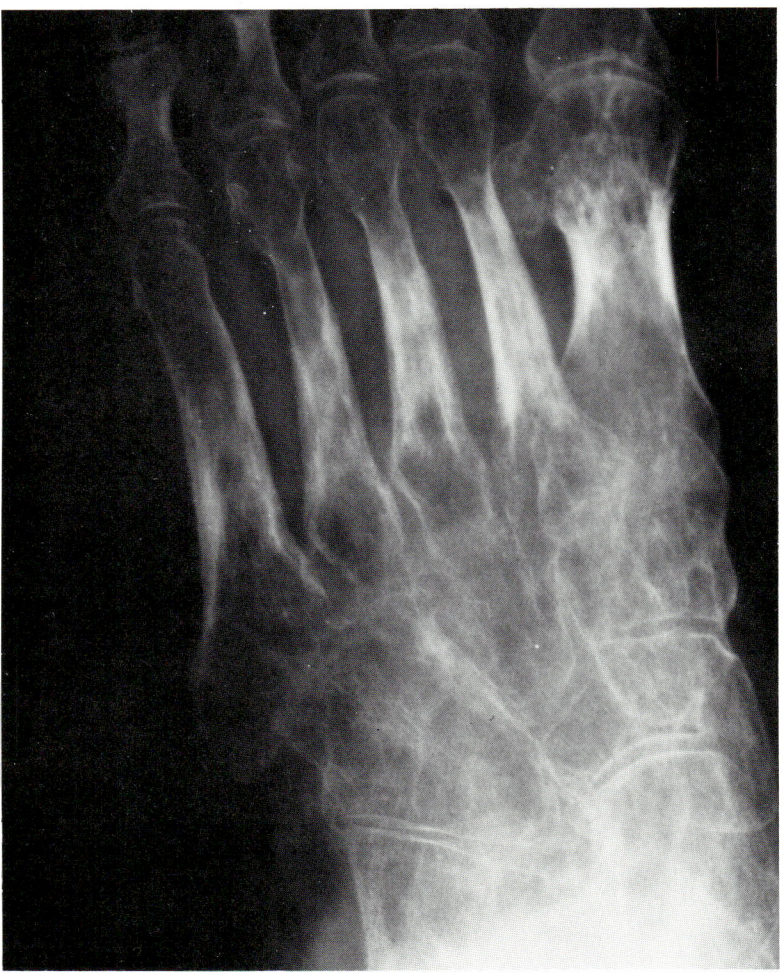

Figure 8.3. Osteoporosis (Sudeck's atrophy) of foot 5 months following fracture of patella. The foot was extremely tender. Brawny edema and cyanosis extended above the malleoli. Two years later, there was a complete recovery. Compare the diminished radiodensity of most bones with the more normal quality in the metatarsal shafts.

changes in the shoulder and hand which restrict motion of these joints, but severe osteoporosis develops in the wrist and metacarpals of the affected side.

By far the commonest variety of osteoporosis is the so-called senile or post-menopausal type. Neither name is felicitous since people do not like to be called senile (especially if they are!) and post-menopausal suggests an unproven etiology. This represents an extension or exaggeration of the age-dependent remodeling of the skeleton. Following the 5th decade of life there is a progressive imbalance between the formation and resorption of bone. The resorption takes place principally on the endosteal surfaces with the result that much trabecular bone disappears and the cortex becomes thinned from the medullary aspect while the outside diameter is preserved or increases. The development of osteoporosis is so slow that excessive osteoclastic activity is not seen in biopsy sections. Because of its endosteal nature, the sites most commonly fractured are the vertebral column and the femoral neck, predominantly cancellous structures. More than 25 percent of women and 15 percent of men above the age 50 years have vertebral compression resulting from osteoporosis.

Spinal fractures are insidious and spontaneous. They occur in small steps but vertebral bodies finally collapse as a result, particularly in their anterior portions. They thus appear wedge-shaped when viewed laterally (Fig. 8.4). Dorsal kyphosis, lumbar lordosis, and a reduction of body height occur. The changes are most marked in the lower thoracic region but no part is spared. Dull pain is the usual complaint and it may radiate in the distribution of the affected spinal roots. Aside from the reduction of radiodensity attending the loss of bone and collapse of the vertebral bodies, the latter frequently become biconcave. The "codfish" appearance (Fig. 8.5) of the vertebrae comes about because the bony cortex next to the intervertebral disc is no longer strong enough to resist the swelling pressure of the nucleus pulposus as discussed in Chapter 1. Localized intrusions of disc tissue and subchondral resorption of bone (Schmorl nodes) are not unduly common in osteoporosis and other osteopenic conditions.

Fractures of the femoral neck increase exponentially following the 5th decade. By the age of 80, 1 of every 10 persons breaks a hip; by 90, 1 of 5. These fractures usually are precipitated by only a trivial trauma. They threaten life but recent advances in orthopedic surgery have made the prognosis less grave. The fractured bone is removed and replaced by metal prostheses.

No unique metabolic abnormality is known to cause senile osteoporosis. It is seen far more frequently in whites than blacks, and in women more than men. The serum levels of Ca, P, and alkaline phosphatase are normal. Three general mechanisms have been considered and each has a therapeutic logic.

1. Endocrine disturbances. It is well known that glucocorticosteroid hormones produce a severe osteoporosis. Administration of these compounds for other conditions in the elderly must reckon with this risk. Because osteoporosis appears so often following the menopause, long-term administration of estrogens has its advocates. It is, however, no magic cure and a deficiency of sex hormones has not been proven to be the basis of senile osteoporosis. One respected authority has suggested that low grade hyperparathyroidism may be the underlying abnormality. The usefulness of calcitonin in treating the condition is being tested clinically.

2. Dietary deficiency. A theory that insufficient intake of calcium by the elderly causes osteoporosis has not stood up but dietary inadequacies should surely be corrected. Recent studies of aged persons have disclosed unexpected instances of osteomalacia arising from nutritional neglect.

3. Disuse atrophy. Senile osteoporosis and physical inactivity re-enforce each other. To break the vicious cycle, judicious exercises and other physiatric measures are an integral part of the management. Analgesics and local anesthetics are the principal measures for control of pain.

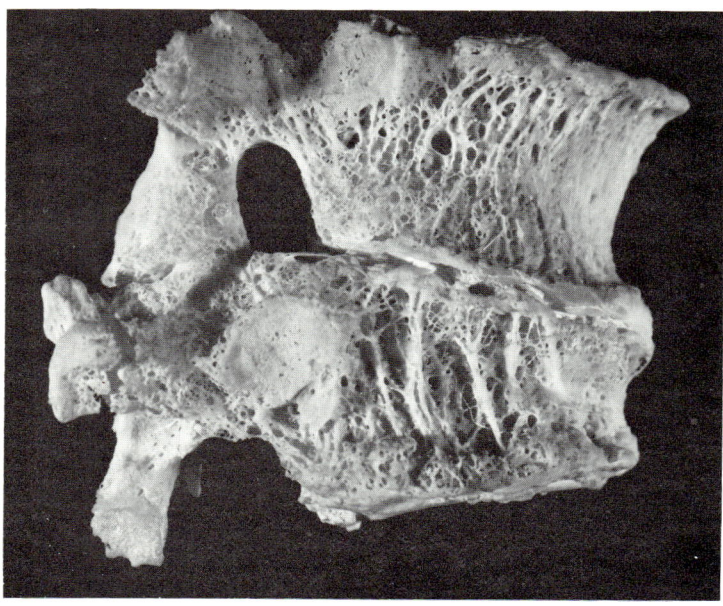

Figure 8.4. Senile osteoporosis, 79-year-old white woman. The cortices of these thoracic vertebrae have been reduced to a delicate filigree and there is a wedge-shaped compression of the lower body.

OSTEOMALACIA

Osteomalacia (literally bone softening) is not a disease but a group of diseases in which a failure of mineralization leads to overproduction of an osteoid matrix (Figs. 8.1 and 8.6). One must infer that in the normal sequence of bone formation, mineral deposition is part of the mechanism for switching off synthesis of the organic matrix. The excessive osteoid is not accompanied by unusual numbers of osteoblasts in most sections

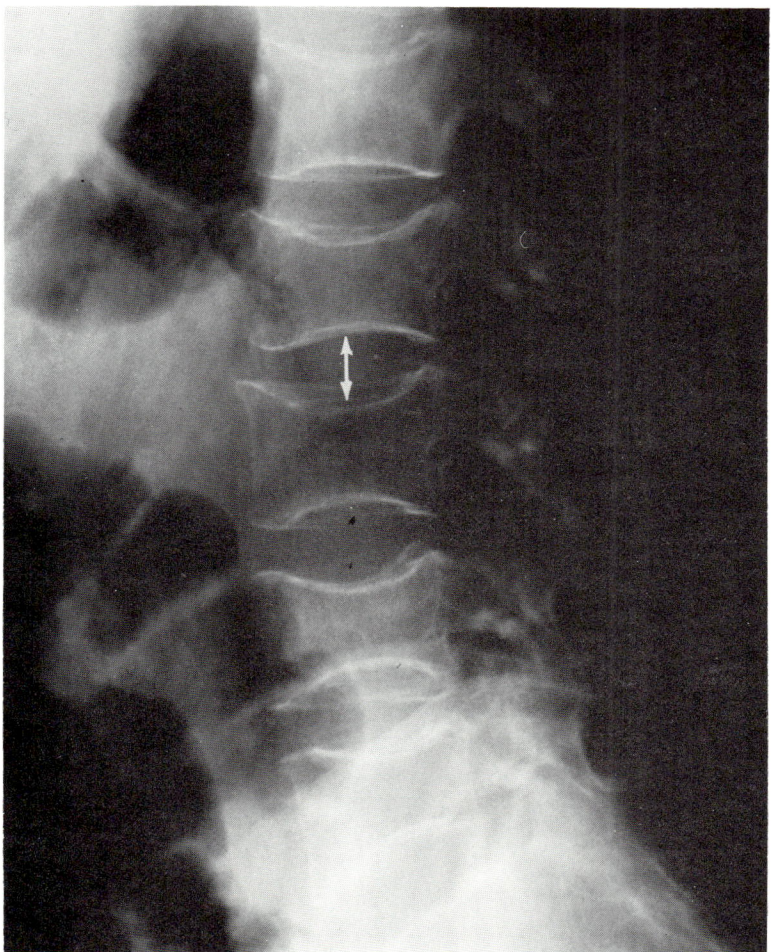

Figure 8.5. Codfish vertebrae in osteoporosis. Note the paucity of trabeculae within the bone. The exaggerated biconcavity of the vertebral bodies has resulted from remodeling of the fragile bone about the turgid intervertebral discs. The latter retain a normal hydrostatic swelling pressure (*arrows*).

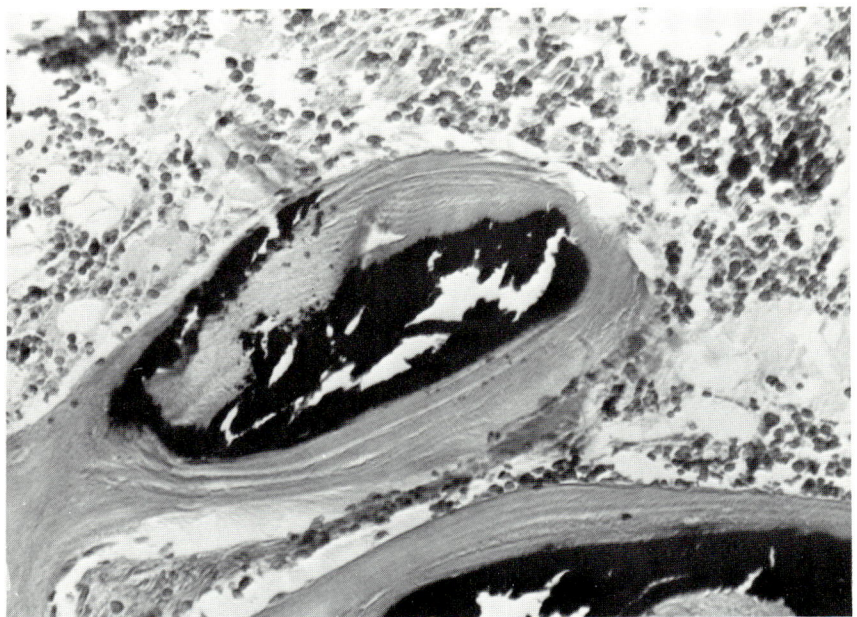

Figure 8.6. Osteomalacia, undecalcified preparation. The black mineralized bone is located centrally within the trabeculae and has been shattered by the microtome knife. The osteoid material adjacent to it appears gray and has a lamellar arrangement. (Von Kossa, ×225.)

because it persists indefinitely once it has been deposited by the cells. It does so because osteoclasts can act only on mineralized matrices. Usually osteomalacia is associated with an absolute reduction in the amount of calcified bone but this is not always the case. In osteomalacia bone tissue has a relatively low mineral content and therefore is soft, *i.e.*, it has a low modulus of elasticity. It differs thus from osteoporosis where bone tissue has a normal mineral content and is not softened. Softening causes osteomalacic bone to become bowed and deformed, while in osteoporosis discrete fractures are more likely. This occurs because osteoporotic bone is fragile rather than soft. Transverse linear defects in osteomalacic bone (Looser lines) resemble fractures with little displacement or external callus but represent localized areas of osteoid replacement rather than true fractures.

In its original sense osteomalacia was a disease of adults resulting from dietary deficiency of vitamin D. Osteomalacia is analogous in every way to rickets, a disorder of children in whom the growth plates are additionally involved. Both conditions may arise from disturbances in the metabolism of calcium, phosphate, and vitamin D. Dietary deficiency of vitamin D no longer is a major factor in the etiology of rickets and osteomalacia in the West, thanks to enrichment of milk, but it remains so in impoverished parts of the world. More common here is the failure of vitamin D, like other fat-soluble vitamins, to be absorbed in intestinal disorders that cause steatorrhea. Formation of insoluble calcium soaps from fatty acids in the chyme also reduces availability of ionic Ca. These malabsorption syndromes may result from celiac disease, non-tropical sprue, chronic pancreatitis, cystic fibrosis of the pancreas, biliary atresia or fistulas, or regional enteritis.

The abnormalities of the epiphyseal plate that characterize rickets result from failure of the cartilage to undergo provisional calcification and hence the sequence of ossification and remodeling. Overgrowth of chondrocytes leads to irregular thickening and widening of the zone (Figs. 8.7 and 8.8). In past generations, severe rickets occurred primarily in infants. The bones became so deformed that there was a reduction of stature approaching dwarfism. Fractures, however, were infrequent. The cranium was misshapen, thinned posteriorly (craniotabes) and thickened in the frontoparietal regions (bossing). Enlargement of the growth cartilage at the costochondral junctions resulted in a chain of knobby protuberances beneath the skin that resembled a necklace—the "rachitic rosary." Dentition was delayed and muscular development was poor.

Rickets and osteomalacia develop through two separate routes in renal disease. Defective resorption of phosphate by renal tubules causes phosphaturia and hypophosphatemia. The latter is the principal basis for poor bone mineralization in one form of renal rickets. The phosphate leak may be an isolated familial defect of the kidney or it may be associated with other tubular dysfunctions, e.g., aminoacidurias (Fanconi syndrome), cystinosis, and renal tubular acidosis. In such instances, the rickets is resistant to vitamin D.

The other renal osteomalacia is superimposed on osteitis fibrosa caused by extensive destruction of renal tissue. Here hypocalcemia is the prime element in the faulty mineralization and it is related to hypovitaminosis D, since the renal lesions have progressed to the point that there is insufficient parenchyma to synthesize $1,25\text{-}(OH)_2D_3$.

In hypophosphatemic osteomalacia, depletion of mineralized bone is less prominent than in the vitamin D-deficient varieties. Indeed an excess of bone, mostly non-mineralized, often occurs in the former. It may obliterate hematopoietic marrow to the point that myelophthisic anemia develops. This is one reason for variable radiologic patterns in osteomalacia. Aside from symmetrical Looser lines, there is rarefaction, i.e., a reduction in the radiopacity of bones. The rarefaction resembles that of osteoporosis but tends to be more diffuse while in osteoporosis those trabeculae which persist appear distinct. In both disorders, codfish deformities and compression of vertebral bodies are seen. Because a

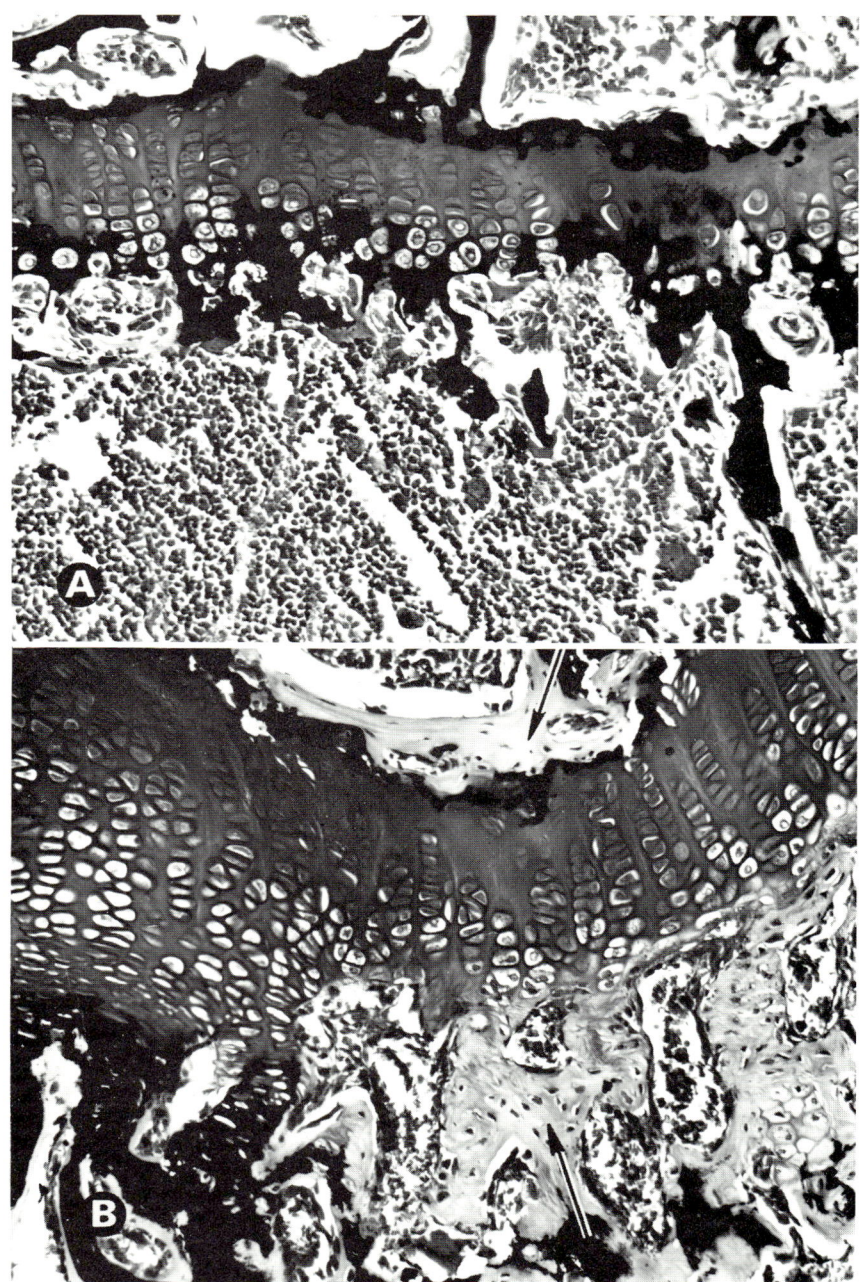

Figure 8.7. Undecalcified sections of experimental vitamin D-deficiency rickets, mouse humerus. *A.* Control. *B.* Rickets. The growth plate is thickened and irregular. There is no black-stained mineral in its basal portion or in the osteoid of the metaphysis or secondary center above (*arrows*). (Von Kossa, ×167).

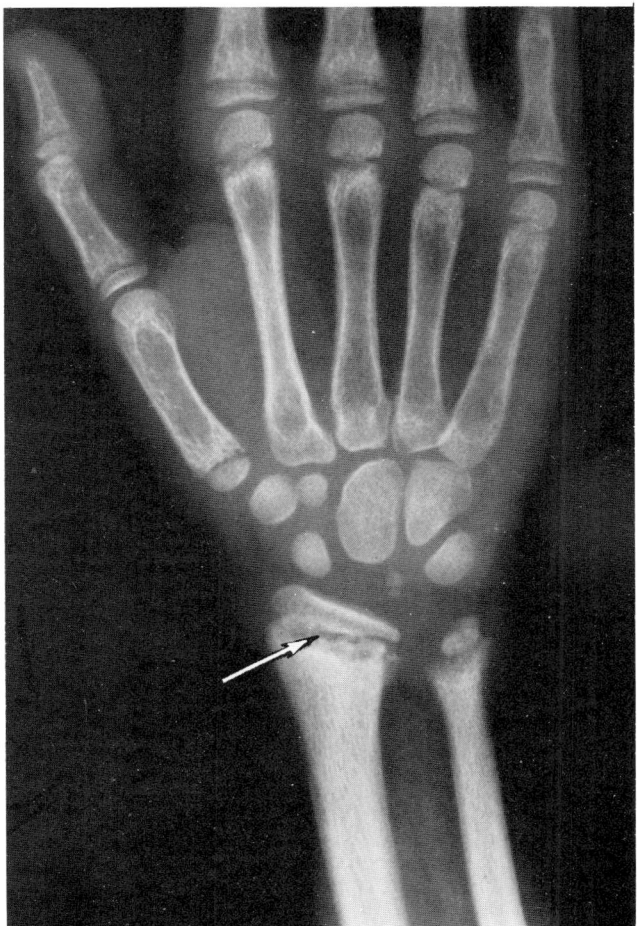

Figure 8.8. Radiologic appearance of rickets in child with Fanconi's syndrome. The two distinctive features are the jagged appearance and excessive "space" between the metaphysis and secondary center of ossification. These correspond to the exuberant growth plate cartilage.

younger population is involved in osteomalacia, the normal swelling pressure of the nucleus pulposus favors development of the codfish vertebrae; this may not be seen in the elderly osteoporotic individual who has degenerated intervertebral discs. In hypophosphatemic osteomalacia, rarefaction is patchier than in the malabsorption syndrome types because the hyperostotic component may be considerable. The roentgenologic hallmark of rickets is the irregularity, thickening, and widening of the growth zone (Fig. 8.8).

Skeletal complaints referable to osteomalacia are similar to those of osteoporosis. Radicular backache and other bone pain may be marked and, in severe cases, femoral neck fractures also occur. If hypocalcemia is present, it may be manifest in increased neuromuscular irritability ranging from positive Chvostek or Trousseau signs to tetany.

The biochemical abnormalities are complex because of the various compensatory homeostatic mechanisms. Common to most of them is an elevated serum alkaline phosphatase activity. In vitamin D-deficiency rickets and osteomalacia, serum levels of Ca frequently are not low because there is a secondary hyperplasia of the parathyroid glands. Serum Ca concentration also is normal in hypophosphatemic rickets, but it decreases in the early azotemic variety as phosphate retention occurs.

The management of osteomalacia varies with its pathogenesis. To the extent possible, the underlying cause should be removed. Correction of vitamin D-deficiencies arrests the rachitic process quite promptly but remodeling is slow and often incomplete. Vitamin D has nothing to offer the hypophosphatemic varieties and, in excessive quantities, may lead to hypercalciuria and renal stones. In malabsorption syndromes, the vitamin is more effective administered parenterally than by mouth. In azotemic varieties a physiologically active derivative of vitamin D should be of therapeutic value but it has yet to be manufactured economically.

OSTEITIS FIBROSA

Osteitis fibrosa is the bony lesion of hyperparathyroidism and is characterized by aggressive osteoclastic resorption of bone (Fig. 8.1). This is accompanied by compensatory osteoblastic activity with some deposition of loose textured endosteal fibrous tissue as well as osteoid. Localized and irregular thinning of trabeculae distinguish it histologically from the preceding disorders (see Fig. 1.7), and these structures often appear gouged out. Small numbers of lymphocytes and mononuclear cells may infiltrate the fibrous tissue but, despite the name, the process is not basically inflammatory. When extensive resorption of bone results in large radiolucent defects, it is called osteitis fibrosa cystica even though true hemorrhagic cysts are rare.

Hyperparathyroidism may be of primary or secondary type, the former arising from functioning tumors, usually adenomas, or primary hyperplasia; and the latter the consequence of renal insufficiency, rickets, or osteomalacia. The clinical picture of hyperparathyroidism classically is "stones, bones, and groans" (nephrolithiasis, osteitis fibrosa, and hypercalcemia). Primary hyperparathyroidism, however, usually presents with kidney stones more than with bone changes. Most osteitis fibrosa is of renal origin and is a particularly conspicuous feature of the patient maintained on chronic hemodialysis. As already indicated, osteomalacia may be superimposed on the osteitis fibrosa and even dominate its appearance. The term renal osteodystrophy subsumes both of these lesions. When young children are affected, the result is often referred to as renal rickets.

The biochemical abnormalities and associated clinical manifestations vary with the underlying process. In primary hyperparathyroidism, serum calcium concentrations are elevated and large amounts are excreted in the urine. Hypophosphatemia develops because parathyroid hormone (PTH) reduces phosphate resorption by the renal tubules. Renal clearance of phosphate constitutes a useful and sensitive test for hyperparathyroidism. Elevated serum levels of PTH can be demonstrated by radioimmunoassay. Because both the calcium and the

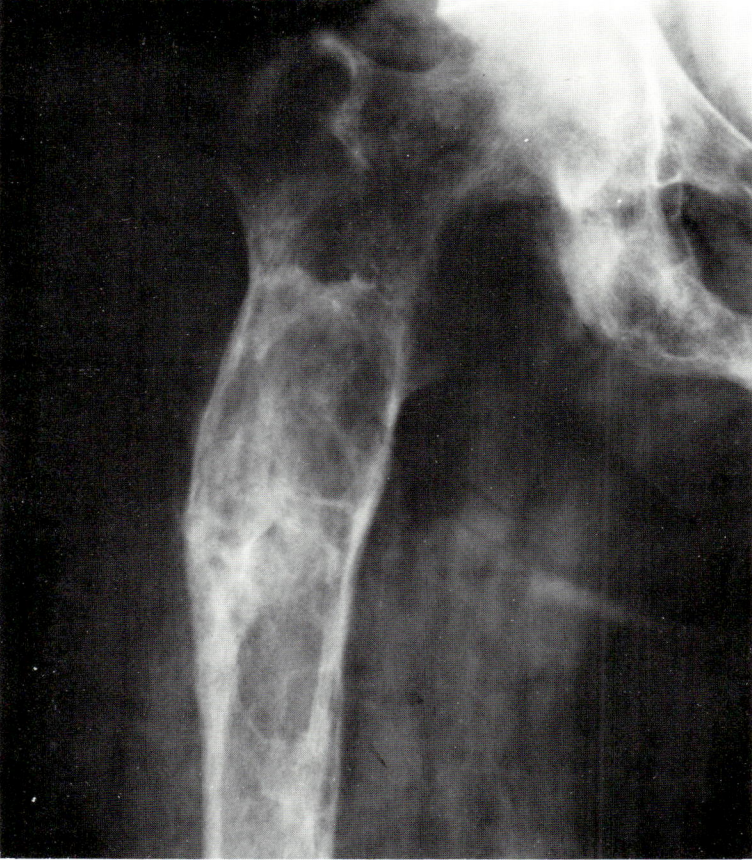

Figure 8.9. Osteitis fibrosa cystica in primary hyperparathyroidism. Cystic rarefaction is seen in the pelvis as well as the femur. Expansion of the cortex in the latter is unusual in this condition and presumably arose from hemorrhage into a "cystic" area.

phosphate levels in the urine are high, calcium phosphate stones are frequent complications. Lethargy, weakness, and vomiting are expressions of the hypercalcemia, and may progress to coma and death.

In renal secondary hyperparathyroidism, the serum electrolyte pattern is quite different. Chronic acidosis is frequent in renal insufficiency and its significance in causing osteodystrophy is debated. Decreased glomerular filtration causes phosphate retention and this is accompanied by a reduction of serum calcium. Most of this Ca remains ionized because of the low pH, and tetany is uncommon. In primary as well as secondary hyperparathyroidism, elevated alkaline phosphatase activity of the serum is characteristic, presumably reflecting the increased synthesis of osteoid by the osteoblasts. The clinical expressions of hyperparathyroidism are influenced by the calcium, phosphate, and vitamin D content of the diet. This may explain the relative infrequency of bone changes in primary hyperparathyroidism and geographic variations in the character of renal osteodystrophy.

The skeletal complaints in osteitis fibrosa are principally pain and tenderness in the extremities and back. The spinal changes may resemble those of osteoporosis and osteomalacia. In addition, there may be pseudocystic rarefactions in x-ray films (Fig. 8.9). Pathologic fractures of the femoral neck and other bones occur in severe instances. Destruction of alveolar bone is an early manifestation and teeth may fall out. Metastatic calcification of periarticular and other soft tissues due to hypercalcemia is frequent and joint cartilage also may be affected.

Resection of the offending glands in primary hyperparathyroidism usually is followed by rapid healing of osteitis fibrosa. In the secondary form, vitamin D therapy may partly reverse the bone lesions but, as in osteomalacia, the risk of nephrolithiasis must be kept in mind.

OSTEOPENIAS SECONDARY TO NEOPLASTIC AND OTHER MEDULLARY LESIONS

Disseminated tumors and other lesions that destroy bone may simulate metabolic bone disease. Metastatic carcinomas usually produce discrete osteolytic defects, but occasionally the bone loss is diffuse and difficult to distinguish roentgenographically from osteomalacia or osteoporosis. Not only does massive osteolysis cause pain and pathologic fractures but sometimes also hypercalcemia and hydroxyprolinuria. More than 7 percent of women with metastatic breast cancer develop symptoms of hypercalcemia and many more the asymptomatic chemical changes. The mechanism for bone destruction in metastatic carcinomatosis is unknown. Osteoclasts are conspicuous by their absence. One possibility is that osteoclasts do the job but then disappear because they are short-lived. Another is that the tumor cells secrete osteolytic substances. Breast cancers usually produce sterols which favor osteolysis and at least one experimental tumor forms prostaglandin E_2. In about one-fourth of patients who have hypercalcemia secondary to breast carcinoma, no grossly detectable changes are seen in bone films.

Elevated serum alkaline phosphatase levels in metastatic bone disease result from reactive new bone formation. In metastatic prostatic carcinomas, and less often in mammary or Hodgkin's disease, the new bone is extensive and the lesions appear more osteosclerotic than osteolytic. Acid phosphatase levels in serum are high in prostatic carcinomatosis because this enzyme is produced by the tumor cells.

Androgenic hormones are employed to promote repair of osteolysis caused by metastatic breast cancer. Synthetic analogues of testosterone have an advantage in retaining the anabolic but reducing the virilizing properties of the hormone. Glucocorticoids are widely employed for reducing hypercalcemia. Sometimes they are very effective, perhaps through acting on the bone, perhaps on the tumor cells. Their long-term use is limited because the same compounds ultimately promote osteoporosis.

Neoplastic and non-neoplastic proliferation of hematopoietic tissues may also cause rarefaction of bone and hypercalcemia. Serum calcium levels are elevated in 70 percent of multiple myeloma patients and less often in those with leukemia. Alkaline phosphatase levels generally are normal. Benign hematopoietic diseases that cause osteopenia include Cooley's (thalassemia) and sickle cell anemias. In both these inherited hemoglobinopathies, destruction of blood is followed by compensatory erythroid hyperplasia of the bone marrow. The hyperplasia results in osteoporosis and may interfere with growth in young children, perhaps by causing pressure atrophy or disturbing the microcirculation. In sickle cell anemia, the microcirculation is further embarrassed by impaction of the red cells. When sufficiently severe, it causes overt bone infarcts. In thalassemia major, growth is restricted but the facies become mongoloid because periosteal bone proliferates at the same time that the spongiosa is thinned out.

The skeleton also is frequently involved secondarily in inherited diseases of lipopolysaccharide metabolism. Unlike the mucopolysaccharidoses, the metabolic problem is not in the osseous tissues but in the bone marrow. In Gaucher's disease, there is a deficiency of an enzyme that can cleave the glucocerebroside, kerasin. Large amounts of kerasin accumulate within reticuloendothelial cells of the bone marrow as well as liver, spleen, and lymph nodes. Gaucher cells are large and have a pale reticulated as distinct from a vacuolated cytoplasm. Most of the cells have a single eccentrically placed nucleus but some have more than one. The lipid cannot be demonstrated by conventional fat stains. Because

Gaucher cells are macrophages, cells which normally have a large lysosomal content, serum levels of acid phosphatase are elevated. Gaucher's disease has a variable clinical course and genetic pattern. Most often it is transmitted as an autosomal recessive and occurs in Jews. A neuronal form causes mental retardation in infants. In other cases, the disorder manifests itself later in life; here hematologic and osseous changes dominate the clinical picture. Bony changes come about because of crowding by the Gaucher cells. Growth may be impaired. A characteristic radiologic finding is the "Erlenmeyer flask" appearance of the lower end of the femur: it is enlarged, the cortex thinned, and the marrow cavity expanded. In adults, the infiltration causes a patchy resorption of bone or aseptic necrosis and so leads to fractures.

REFERENCES

BARTTER, F. C. Bone as a target organ; toward a better definition of osteoporosis. Perspect. Biol. Med. 16:215, 1973.

BARZEL, U. S. (Ed.) Osteoporosis. New York, Grune & Stratton, 1970.

BISHOP, M. C., WOODS, C. G., OLIVER, D. O., LEDINGHAM, J. G. G., SMITH, R., AND TIBBUTT, D. A. Effects of haemodialysis on bone in chronic renal failure. Br. Med. J. 3: 664, 1972.

DAVIS, H. L., JR., WISELEY, A. N., RAMIREZ, G., AND ANSFIELD, F. J. Hypercalcemia complicating breast cancer. Clinical features and management. Oncology 28:126, 1973.

DEITRICK. J. E., WHEDON, G. D., AND SCHORR. E. Effects of immobilization upon various metabolic and physiologic functions of normal men. Am. J. Med. 4:3, 1948.

DOTY, S. B., AND MATHEWS, R. S. Electron microscopic and histochemical investigation of osteogenesis imperfecta tarda. Clin. Orthop. 80:191, 1971.

GARNER, A., AND BALL, J. Quantitative observations on mineralized and unmineralized bone in chronic renal azotaemia and intestinal malabsorption syndrome. J. Pathol. Bacteriol. 91:545, 1966.

LENNON, E. J. Metabolic acidosis. A factor in the pathogenesis of azotemic osteodystrophy? Arch. Intern. Med. 124:557, 1969.

MALLETTE, L. E., BILEZIKIAN, J. P., HEATH, D. A., AND AURBACH, G. H. D. Primary hyperparathyroidism; clinical and biochemical features. Medicine 53:127, 1974.

MANKIN, H. J. Rickets, osteomalacia and renal osteodystrophy. J. Bone Joint Surg. 56A:101, 1974.

MOSELEY, J. E. Bone Changes in Hematologic Disorders. (Roentgen Aspects.) New York, Grune & Stratton, 1963.

NEWTON-JOHN, H. F., AND MORGAN, D. B. The loss of bone with age, osteoporosis and fractures. Clin. Orthop. 71:229, 1970.

POWELL, D., SINGER, F. R., MURRAY, T. M., MINKIN, C., AND POTTS, J. T., JR. Nonparathyroid humoral hypercalcemia in patients with neoplastic diseases. N. Engl. J. Med. 289: 176, 1973.

9

HYPEROSTOTIC DISORDERS

Diseases in which the quantity of mineralized bone is excessive are much less frequent than those in which it is reduced. The exceptions are Paget's disease and the osteoblastic metastases already described.

PAGET'S DISEASE OF BONE

This is an idiopathic, slowly progressive condition in which bone becomes deformed and sclerotic through disorderly resorption and apposition of new osseous tissue (Fig. 9.1). It may affect one or many bones but always is a localized rather than a diffuse process. Paget's disease is by far the most common of the hyperostoses although there are marked geographic variations in its occurrence. In England, up to 15 percent of the adult population at necropsy has the lesion, usually in monostotic form. Clinically significant polyostotic disease is much less frequent. It rarely occurs in Chinese.

Paget's disease most often becomes apparent during middle-age and is rare before 20 years. Often it is discovered accidentally in x-ray films taken for other purposes. The calvarium, vertebral column, pelvis, and long bones are the sites of predilection. Dull pain may be the presenting complaint but more often it is deformity such as enlargement of the head (Fig. 9.2). Fractures are frequent and the lower extremities become bowed. The disease, as indicated by serial radiologic examination, begins at one end of the bone and progresses toward the other. Early it appears as a wedge-shaped rarefaction of the cortex, the apex pointed toward the center of the bone. Histologically this corresponds to active osteoclastic resorption. In the wake of the advancing resorptive process, this phase is followed by deposition of new bone, mostly periosteal. The diameter of the bone is widened thereby. The chaotic new bone formation and resorption seemingly violate Wolff's law and show up microscopically as a mosaic pattern of cement lines (Fig. 9.3), a useful diagnostic feature in biopsies. The intertrabecular spaces are occupied by fibrous tissue which in the active phases contains numerous large, thin-walled blood vessels. A vascular bruit is sometimes heard when a stethoscope is placed over

98   *Musculoskeletal System*

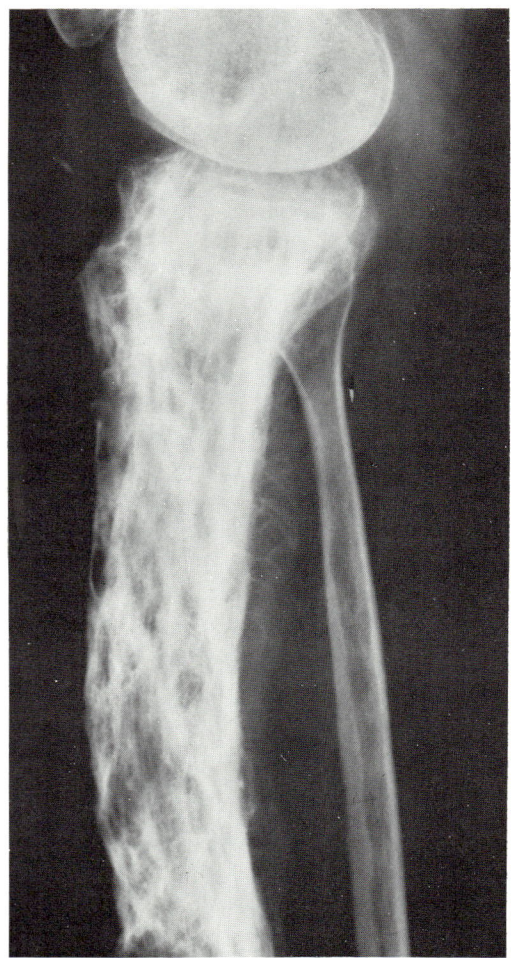

Figure 9.1. Advanced Paget's disease. The tibia is bowed anteriorly and contains a great excess of disorderly bony tissue. The diameter of the fibula as well as the tibia is increased.

the bone. These features have long been considered evidence for a vascular basis of Paget's disease. No arteriovenous anastomoses have been found. High output cardiac failure occasionally results in association with this large vascular bed.

The cause of Paget's disease is unknown. There is no evidence that a specific metabolic defect of bone is at fault. The serum Ca and P levels usually are normal and the outstanding chemical abnormality is marked elevation of alkaline phosphatase.

Deafness may be severe and results from impingement of the thick-

Hyperostotic Disorders 99

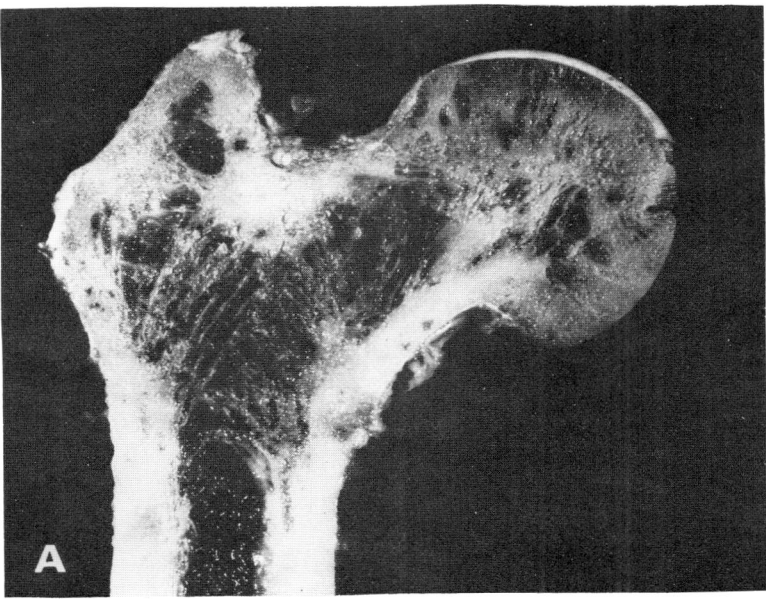

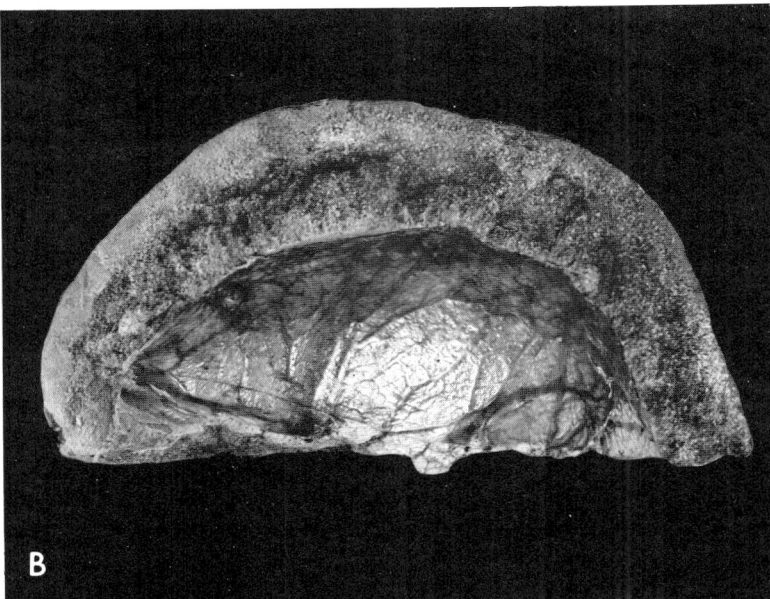

Figure 9.2. Paget's disease. *A.* The cortex of the shaft is irregularly thickened and the cancellous bone in the femoral head is variously sclerotic and porotic. The neck of the femur has remodeled so that it is unusually horizontal with respect to the shaft. *B.* Massive thickening of the calvarium.

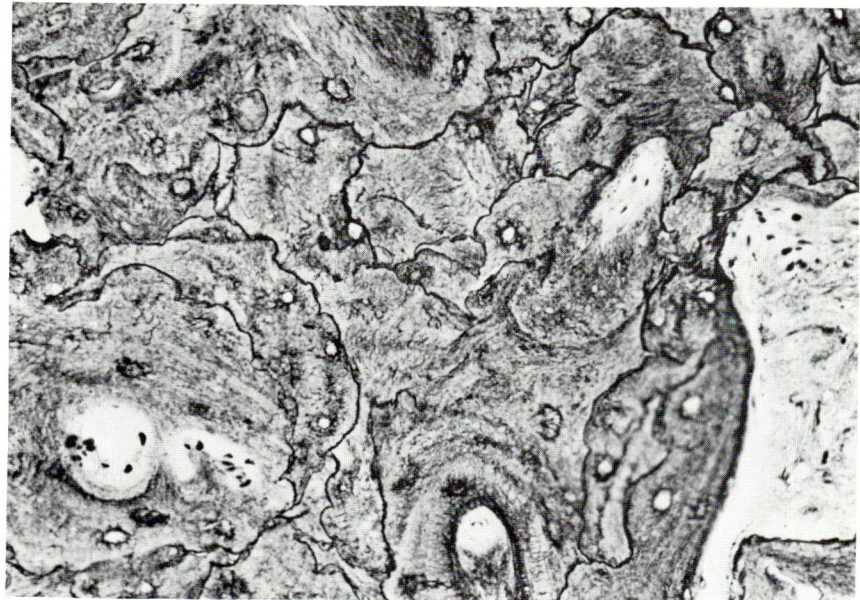

Figure 9.3. Mosaic pattern of cement lines in Paget's disease. Compare with normal in Figure 1.5. (Bodian's protargol, ×225.)

ened temporal bone on the auditory nerve. Approximately 2 percent of patients with Paget's disease develop osteogenic sarcomas. Most if not all osteogenic sarcomas in the elderly arise in this manner, even if the underlying disease was not recognized previously.

There is no specific treatment for this condition. Calcitonin is being tried during the active phases but it is too early to evaluate its worth.

OSTEOPETROSIS

This condition, also known as marble bone or Albers-Schönberg disease, is a relatively rare inherited disorder in which the skeleton is excessively compact and mineralized (Fig. 9.4). The size and the shape of the bones are normal. In most instances, it is present at birth and death occurs in a few months. It appears also in a more benign tardive form later in life. The basic abnormality is a failure of the mechanism for removing calcified cartilage during endochondral ossification (Fig. 9.5). Cartilage normally is more heavily calcified than bone. In osteopetrosis, unusually large osteoclasts may be present but they apparently are ineffective. The bones, despite their great hardness, are brittle. Whether this is due to loss of viscoelasticity or stress concentration along the cartilage-bone interfaces is unknown. Fractures are common and may be

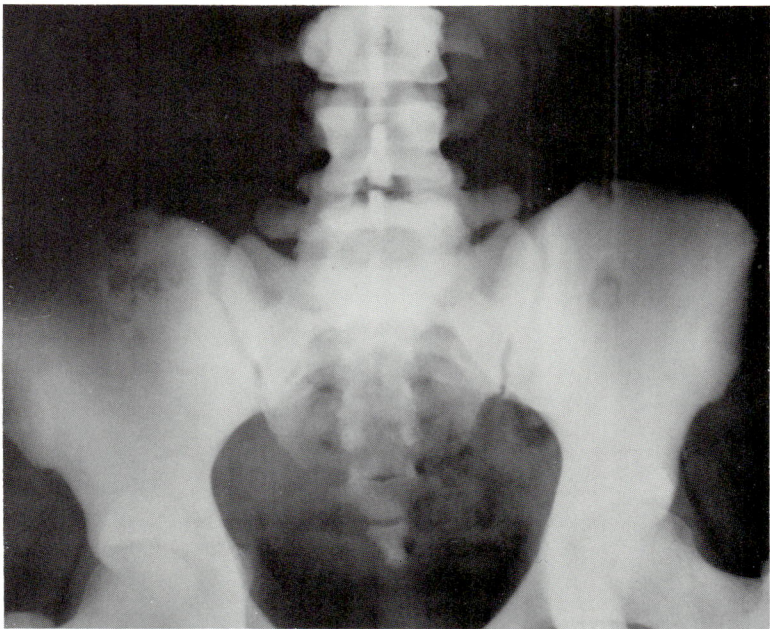

Figure 9.4. Radiopacity of spine and pelvis in a 21-year-old woman with asymptomatic osteopetrosis. The bone appears excessively white but has a normal configuration.

the first evidence of the disorder in infants. The serum concentrations of Ca, P, and alkaline phosphatase are normal. Aside from fractures, the principal consequences of osteopetrosis are myelophthisic anemia and its concomitants.

HYPERTROPHIC OSTEOARTHROPATHY

Secondary or pulmonary hypertrophic osteoarthropathy is an occasional complication of primary intrathoracic tumors or suppuration. It occurs in 5 to 10 percent of cases of bronchiogenic carcinoma and even more of the rarer pleural mesothelioma. There are two main components to the syndrome: clubbing of the fingers and periosteal new bone formation along the long and short tubular bones of the extremities. Clubbing of the fingers alone is a frequent finding in many other circumstances, particularly cyanotic heart disease. It results from thickening of the fibrous tissue of the nail bed. The periosteal new bone, however, is not characteristic of cyanotic heart disease. It is responsible for pain, tenderness, and sometimes swelling of the affected areas. Joint effusions also may occur and, particularly when the primary tumor is occult, cause confusion with rheumatoid arthritis. The pathogenesis of

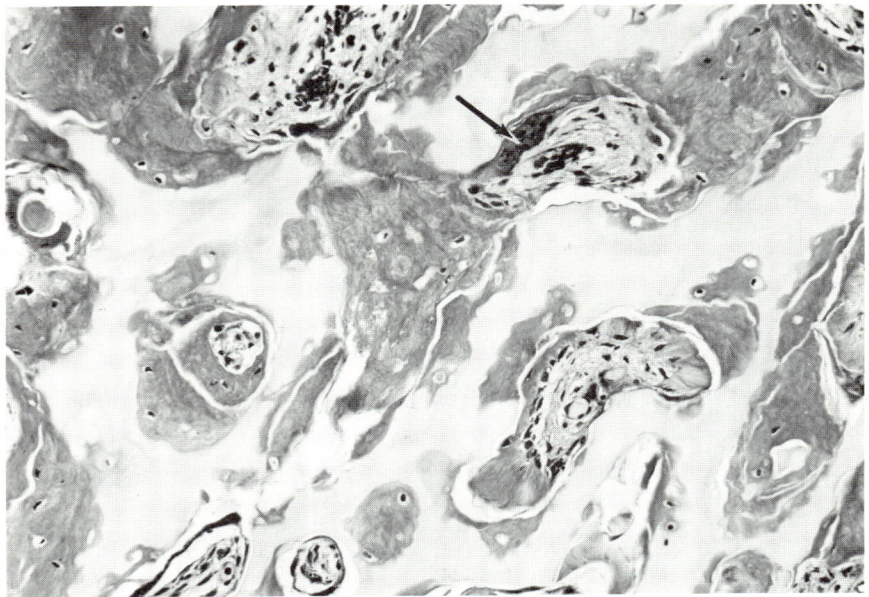

Figure 9.5. Persistence of calcified cartilage in osteopetrosis. The cartilage is the pale acellular material. The darker tissue is woven bone. Note the large osteoclast (*arrow*). (Hematoxylin and eosin, ×160.)

secondary hypertrophic osteoarthropathy is somewhat of a mystery. It seems to involve a vagal reflex action on the periosteal blood flow. Sectioning the vagus nerve or removing the tumor may alleviate pain promptly and cause the periosteal new bone to regress.

Primary hypertrophic osteoarthropathy is a rare familial disease which has no underlying neoplastic or infectious substrate. It usually appears after puberty. The skeletal changes resemble the preceding but there is, in addition, marked thickening of the dermis called pachydermia (elephant skin).

MYELOSCLEROSIS

Agnogenic myeloid metaplasia is one of many terms describing an uncommon idiopathic failure of the bone marrow which results in massive extramedullary hematopoiesis and appearance in the peripheral circulation of immature red and white blood cells. It simulates leukemia and occurs principally in older persons. The cause of the marrow failure is obscure; the best recognized etiologic factor, and this in only a proportion of cases, is exposure to organic solvents such as benzene. The cellularity of the marrow varies greatly, depending on the stage of the disease, but in the end there often is marked fibrous replacement of hematopoietic tissue. Because of the age group affected, those parts of the skeleton having a red marrow—the vertebrae, rib cage, and pelvis—are most likely to be involved. Ossific metaplasia of the fibrous tissue occurs in varying degrees and may cause these bones to appear highly sclerotic in x-ray films. The cortices are unaffected and the contours of the bones are preserved. Complaints are referable to the hematologic findings and discomforts of the bone ordinarily are not important.

CAFFEY'S DISEASE

Infantile cortical hyperostosis is a proliferative periosteitis that occurs in the early months of life. It is uncommon and the cause is unknown. Low grade fever is present, suggesting that the condition is infectious. The mandible is the bone most often involved but the changes may take place throughout the body. Soft tissues about the bone may also be swollen and there is much formation of new bone in the periosteum (Fig. 9.6). Corticosteroids are effective in controlling the condition and the

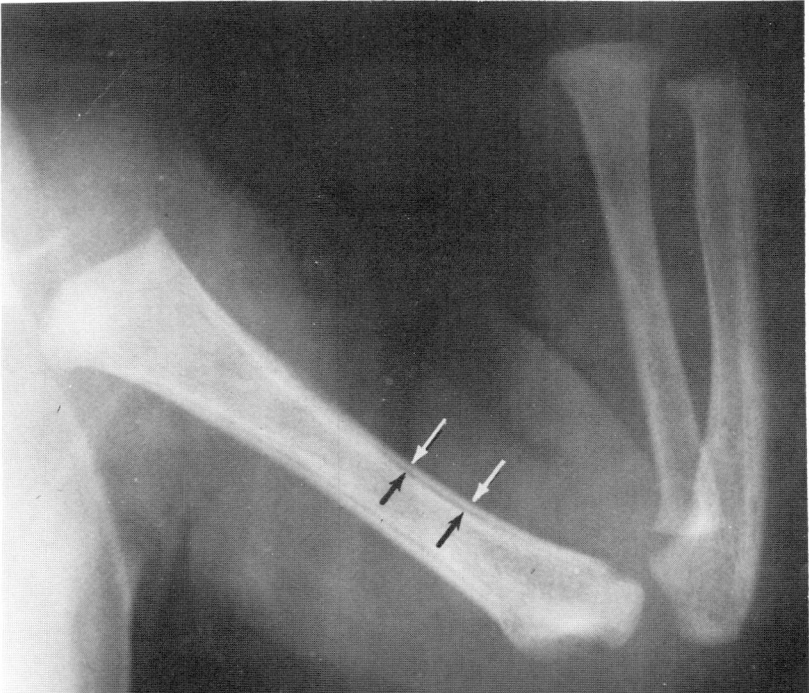

Figure 9.6. Caffey's disease. The appearance of the radiopaque lamina (*arrows*) external to the shaft of the humerus is characteristic of periosteal new bone formation.

appearance of the bone is usually restored within several months with or without treatment.

OSTEOFLUOROSIS

The toxic consequences of fluorosis are most apparent in the teeth. In regions where drinking water contains 1.5 to 3.9 parts per million of fluoride, the enamel of a large part of the population has a mottled brown and white appearance. Degeneration of ameloblasts results in defective mineralization of the enamel. A still higher level of fluoride in the water or industrial exposures may cause marked osteosclerosis. In some manner, the fluoroapatite stimulates proliferation of new bony trabeculae. Most of the matrix is normal morphologically, but there are some woven areas. In addition, unidentified bleb-like deposits sometimes are prominent in such tissues. Mineralization of articular ligaments causes stiffness, particularly of the spine, in about 20 percent of cases but otherwise there are no skeletal complaints.

REFERENCES

BARRY, H. C. Paget's Disease of Bone. Edinburgh, E. and S. Livingstone, 1969.

JOHNSTON, C. C., JR., LAVY, N., LORD, T., VELLIOS, F., MERRITT, A. D., AND DEISS, W. P., JR., Osteopetrosis. A clinical, genetic, metabolic and morphologic study of the dominantly inherited, benign form. Medicine 47:149, 1968.

McCLURE, F. J. Water Fluoridation. The Search and the Victory. Bethesda, Md., National Institute of Dental Research, 1970.

NAGANT DE DEUXCHAISNES, C., AND KRANE, S. M. Paget's disease of bone; clinical and metabolic observations. Medicine 43:233, 1964.

RHODES, B. A., GREYSON, N. D., HAMILTON, C. R., JR., WHITE, R. I., JR., GIARGIANA, F. A., JR., AND WAGNER, H. N., JR. Absence of anatomic arteriovenous shunts in Paget's disease of bone. N. Engl. J. Med. 287:686, 1972.

10

INFECTIONS OF BONES AND JOINTS

Although their frequency has decreased greatly since the advent of antibiotic therapy, infections of bones still account for more than 1 percent of admissions to general hospitals in the United States. The formerly high mortality of suppurative osteomyelitis has been reduced greatly but the morbidity persists. As the name suggests, the seat of the inflammation in osteomyelitis is the bone and its marrow. Osteitis, strictly speaking, means inflammation of bone as distinct from the marrow but both tissues are involved in osteomyelitis and osteitis. In common parlance, osteomyelitis refers to infectious, and osteitis to non-infectious conditions. Indeed most diseases called osteitis, *e.g.*, radiation osteitis, osteitis fibrosa cystica, and osteitis deformans (Paget's disease), are not even inflammatory. Spondylitis is the generic term for spinal inflammation and vertebral osteomyelitis is often called infectious spondylitis (Fig. 10.1). When the inflammation affects the metaphyseal junction of the bone and cartilage, it is called osteochondritis.

OSTEOMYELITIS

Infectious agents gain access to the bone marrow either through the blood stream or by direct inoculation. Examples of the latter are penetrating injuries such as compound fractures, or extension from adjacent soft tissue lesions such as gangrene or infected skin ulcers. The hematogenous forms occur preponderantly in young children but 80 percent of osteomyelitis nowadays is not blood-borne.

The basis for the hematogenous localization of infection in the bone marrow of young children has already been indicated. The terminal ramifications of the nutrient artery make a sharp loop at the metaphysis and enter a system of large sinusoidal channels. The sluggishness of the circulation here and absence of anastomoses favor impaction of bacteria. *Staphylococcus aureus* is the organism in 85 percent or more of such cases. It enters the blood from a remote source. The onset of acute hematogenous osteomyelitis classically is explosive. Some days after another infection—sometimes it is quite minor, such as a carbuncle or

Figure 10.1. Suppurative hematogenous osteomyelitis of lumbar vertebral body. The patient was a 38-year-old man who had been treated with corticosteroid agents for psoriatic arthritis over a 9-month period. The liquefied bone has discharged into the peritoneal cavity through the rent in the prevertebral fascia above. The adjacent intervertebral discs have been spared and are closely approximated.

sinusitis—fever, chills, and pain localized over the femur or tibia appear suddenly. As purulent exudate develops in the marrow, bony trabeculae as well as hematopoietic tissue become necrotic. The exudate soon penetrates through soft tissue gaps in the cortex (Volkmann canals and the periosteal ring) to inflame and elevate the periosteum. Bacteria seed back into the circulation through medullary veins. Thus both blood and subperiosteal exudate are candidates for culture. Radiologic changes do not appear until a week or more has elapsed. Thereafter two characteristic abnormalities are seen. Irregular areas of radiolucence appear at the affected metaphysis. These are mediated largely by osteoclasts in the viable adjacent bone. A mass of opaque, dead bone (sequestrum) may persist in radiolucent areas of suppuration. The other abnormality is the formation of new bone in the cambium layer of the periosteum. The shell of periosteal new bone is the involucrum (Fig. 10.2). Although the involucrum serves largely to wall off the infection, it also renders bacteria persisting in the sequestrum inaccessible to antibiotics. It thus forms a constant source of reinfection and sets the stage for chronic low grade osteomyelitis. Sinus tracts to the skin may form by direct extension and become chronic.

From the above description, three cardinal features of the management of acute osteomyelitis follow: (1) the etiologic agent must be identified promptly, (2) large doses of the appropriate antibiotic admin-

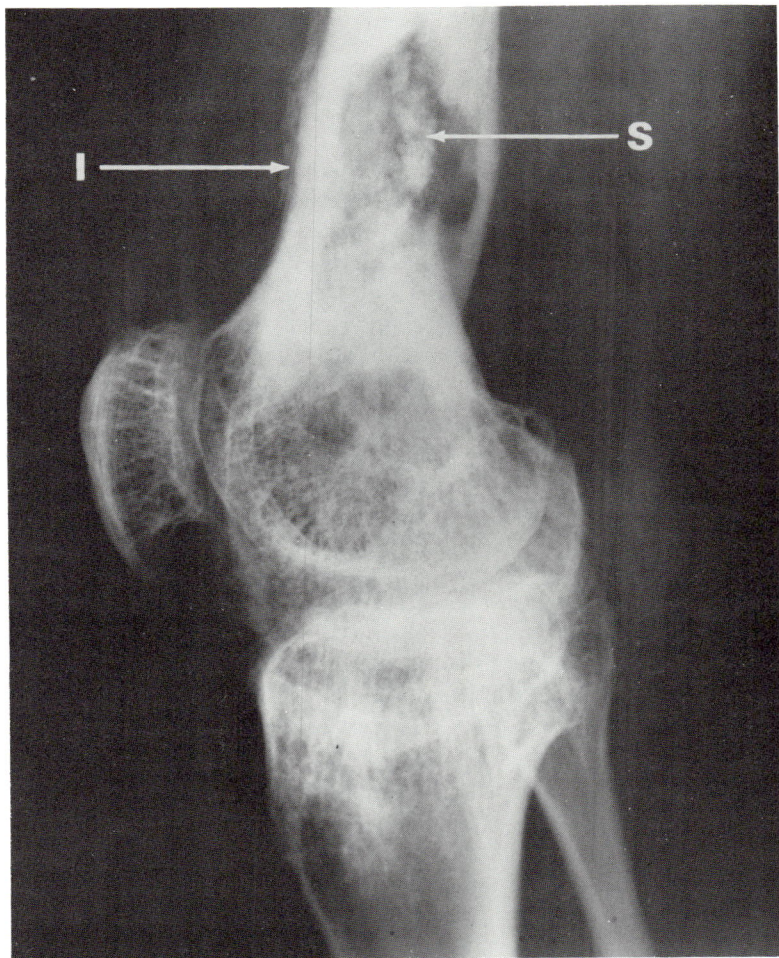

Figure 10.2. Chronic osteomyelitis, distal femur. The sequestrum (*S*) lies within a lucent area of bone destruction and the entire lesion is enveloped by the thick involucrum (*I*).

stered systemically, and (3) surgical drainage. Early bacteriologic diagnosis often is established by culture of the blood or periosteal exudate; the latter may also be examined for bacteria by direct smear. If these measures are insufficient, biopsy of the bone marrow may be necessary.

Other organisms also cause osteomyelitis. Tuberculosis remains a problem despite the fact that bovine infection, traditionally the main cause of skeletal disease, has largely been eradicated. The vertebral column, hip, and knee, in that order are the principal sites affected. The disease is hematogenous and, more often than not, independent of pulmonary or other visceral tuberculosis. Tuberculous spondylitis, Pott's disease, occurs in the lower thoracic and upper lumbar regions. The usual histologic features of tuberculosis in other organs are seen here. In addition to tubercle formation and caseation, there is much collapse of the vertebral bone. The intervertebral discs are relatively less affected because they are avascular and the exudate, being non-suppurative, lacks collagenases. Collapse of the vertebral bodies may proceed to the formation of a kyphos. Accumulation of exudate in the epidural space causes neurologic symptoms but tuberculous meningitis is rare. In half the patients, the inflammatory process burrows along paraspinal tissues and presents at a distance. Thus the tuberculous "psoas abscess" appears at the groin.

In adults, isolated hematogenous vertebral osteomyelitis is increasingly recognized (Fig. 10.1). Systemic complaints may not be obtrusive and the lesion comes to attention because it produces pain and radiologic changes. Brodie's abscess is a post-infectious cyst-like rarefaction of the metaphysis surrounded by a rim of sclerotic bone. It was originally described as a feature of typhoid osteomyelitis but is also produced by other infectious agents. *Brucella suis* should also be considered in the differential diagnosis of spondylitis in endemic areas of brucellosis. Diabetes mellitus and systemic corticosteroid therapy predispose to bone infections. There is a predilection for salmonella organisms to cause osteomyelitis in children with sickle cell anemia; the reason for this is not apparent.

Chronic osteomyelitis with sinus tract formation often is a discouraging condition. Even after they appear to have healed for years, the tracts are likely to open spontaneously or after mild injury, and discharge fragments of sequestrum and exudate. A squamous cell carcinoma develops in the skin of somewhat less than 1 percent of these sinuses after many years.

MISCELLANEOUS BONE INFECTIONS

Syphilis of bone and joint takes many forms. There still occur more than 2000 cases of congenital syphilis each year in the Unites States. In 5 percent of these infants, syphilitic osteochondritis develops. Transplacental infection takes place after the 5th month of gestation. Clinical manifestations therefore are apparent in the first weeks of life and rarely develop after 3 months. As in other hematogenous infections, the treponemata lodge at the metaphyses. They cause necrosis of bone and the osteochondral junction appears jagged as a result. The inflammation also spreads to the periosteum and causes local tenderness. Involvement of the proximal epiphysis of the humerus restricts use of the arm; this is Parrot's pseudoparalysis. Untreated, the epiphysis may fracture or become deformed and fail to grow. Early treatment with penicillin permits complete recovery.

The two varieties of tertiary syphilis of bone, gummatous osteomyelitis and diffuse periosteitis, have now largely passed into history. Rather similar changes also occurred in young children with inadequately treated congenital syphilis. Destruction of the nasal bones led to a saddle nose, and proliferative periosteitis to saber shin deformities. Hutchinson's teeth, small notched incisors, often were associated with these. They presumably developed because spirochetes infected the tooth germ. Yaws, once the commonest form of infectious bone disease, resembled syphilis in many ways. The causative agent, *Treponema pertenue*, is morphologically indistinguishable from *Treponema pallidum* but is not transmitted by sexual contact. The virtual elimination of this formerly widespread tropical disease by penicillin represents a major accomplishment of modern medicine.

The role of viruses in causing congenital anomalies of bone has been demonstrated most convincingly in the case of rubella. Transmission of the virus from mother to fetus during the first trimester of pregnancy causes the agent to appear in most tissues. The virus disappears from most of these tissues as antibodies appear. It persists in cartilage, however, because the matrix excludes these antibodies from the chondrocytes. It is only after the sequences of endochondral ossification replace the chondrocytes that the infection is overcome. Metaphyseal abnormalities are a consequence in some cases. The deafness so common in prenatal rubella, however, is usually of sensorineural rather than bony type.

The animal parasite most commonly affecting the skeleton is the larval dog tapeworm, *Echinococcus granulosus*. It causes slow focal destruction of bone in about 2 percent of cases of echinococcosis. The pelvis and vertebral column are the sites most involved. X-ray examination shows the lesions to be cystic. The condition is seen most often in immigrants from Mediterranean countries. Eosinophilia is present in about a fourth of cases. There is a specific intradermal reaction to hydatid cyst fluid.

INFECTIOUS ARTHRITIS

Joints become infected by three routes: the blood stream, direct inoculation, or extension from an osteomyelitic focus. Direct inoculation is illustrated by infection following surgical procedures on joints (Fig. 10.3). This occurs in approximately 1 percent of total hip replacement

110   Musculoskeletal System

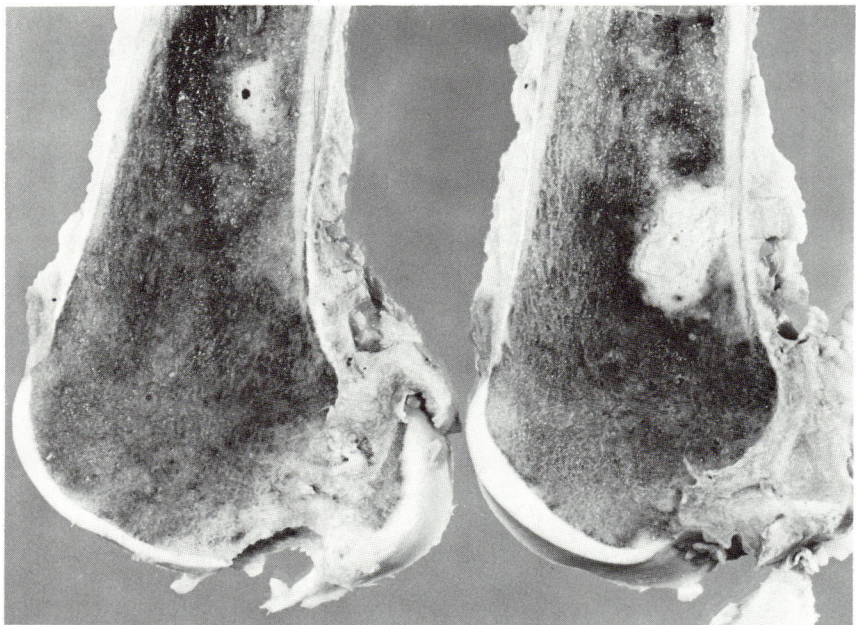

Figure 10.3. Inoculative infectious (staphylococcal) arthritis and osteomyelitis following joint surgery. The marrow adjacent to the defect in the femoral condyle is infiltrated by pale purulent material and an abscess is present more proximally.

procedures under the best of circumstances. Many categories of infectious agents in addition to bacteria—viruses, fungi, and even a few parasites, *e.g.*, the guinea worm, *Dracunculus*—may enter joints. Bacteremias often lead to infectious arthritis. A seeming irony is that vaccination and other antibacterial procedures sometimes increase the occurrence of infectious arthritis in bacteremic states. The vaccination may preserve life and so allow infectious arthritis to come into being rather than itself being an immunopathogenetic factor. If acute meningococcemia does not terminate in the fulminant Waterhouse-Friedrichsen syndrome, meningococcal arthritis is likely to result. This phenomenon should be remembered in interpreting experimental and clinical data on the part hypersensitivity plays in causing arthritis following immunization procedures.

The affinity of microorganisms for joints is only partly understood. Because hematogenous arthritis itself is often secondary to blood-borne osteomyelitis, the circulatory peculiarities of the bone account for part of the arthrotropism. This cannot explain why bacteria which infect synovial tissue in the first place or viruses, such as rubella, lodge in joints. Low concentrations of complement or other local chemical factors must be involved. Phagocytosis by polymorphonuclear leukocytes is less efficient in joint fluid than on rough surfaces. Synovial mucin is not known to influence this. Antecedent rheumatoid arthritis or other joint disease predisposes to seeding of bacteria from the blood.

The pathologic processes in diarthrodial joints resemble those produced by the same organisms in soft tissues with two exceptions: (1) because articular cartilage is avascular, it is more resistant to bacterial invasion than the bone marrow or synovial tissue; and (2) the pluripotentiality of articular granulation tissue leads not only to scar tissue but to a heterogeneous cartilage and bone. Infectious arthritis usually affects one or only a few joints. The large joints, particularly the knee and hip, are most likely to be involved. Exudation into the joint space accompanies the synovial inflammation and periarticular soft tissues are congested and edematous. Periosteal new bone at the margins of joints also follows. Regional lymphadenitis is common and, particularly in granulomatous arthritis, may be a preferred site for biopsy rather than the joint because the risk of fistula-formation is reduced. In purulent arthritides, polymorphonuclear leukocyte enzymes can digest both collagen and ground substance proteins at neutral pH and thus destroy joint cartilage rapidly. By contrast, in granulomatous joint diseases, these enzymes are absent and articular cartilage can persist long after other joint tissues have disappeared. Surgeons have repeatedly observed cartilage floating unattached in the exudate of tuberculous joints.

Arthrocentesis (tapping the joint fluid) is a diagnostic necessity in pyogenic arthritis. The fluid is turbid and has low viscosity. Microscopic examination reveals a large number of cells, 75,000 per mm^3 or more, most of them neutrophils. Bacteria may be seen within them. Microorganisms often are absent from synovial fluid while present in synovial tissue, and the latter should be cultured as well as the fluid during surgical intervention or biopsy of infected joints. A poor mucin clot is formed after addition of dilute acetic acid. The glucose concentration is reduced at least 50 percent below that of blood.

Within the past decade there has been a recrudescence of gonococcal arthritis and this now constitutes the most frequent cause of infectious joint disease. It is an acute febrile illness and appears within a few days to 2 weeks of the venereal infection. The organism is usually seen in urethral or cervical exudate but oropharyngeal and rectal lesions are also common. Within 2 weeks, demineralization of para-articular bone is seen in x-ray films and then destruction of cartilage as indicated by narrowing of the "joint space." The end result may be extensive deformity of the joint and loss of mobility. Early treatment with penicillin as well as removal of the synovial fluid by arthrocentesis, however, usually effects a good cure. Bacterial arthritis is a medical emergency. Delay of optimum therapy for even a few days can have tragic, life-time consequences.

Another venereal arthritis can be confused with gonococcal infection: Reiter's disease. This condition has three main features: non-specific urethritis, conjunctivitis, and arthritis. Venereal transmission is established in most cases, but the causative agent has not been identified. Fever is present in most cases and the attack usually runs its course in several weeks or months. It is subject to recurrence, and limitation of motion or instability of the joint sometimes results. As one would expect in infectious arthritis, periosteal new bone is seen at attachments of articular ligaments and tendons.

In staphylococcal and other bacterial arthritides of children, radiolucent areas of osteomyelitis are seen at the metaphysis and this is useful in distinguishing infectious from juvenile rheumatoid arthritis. Several filtrable agents (*Mycoplasma* and *Chlamydia* sp.) cause epizootic arthritis in other species, avian and mammalian. Despite intensive investigation, comparable organisms have not been found in human joints.

The principal granulomatous infections of joints are tuberculosis and coccidioidomycosis. The two diseases resemble each other and both arise by hematogenous dissemination. Joint involvement usually is secondary to osteomyelitis but some tuberculosis begins in the synovium. These arthritides begin slowly and are not accompanied by redness or fever. Coccidioidomycosis is endemic in the southwestern states and parts of Latin America. Roughly one-third of primary pulmonary infections by the fungus, although self-limited, are accompanied by transient pain and tenderness in joints. Fever and erythema nodosum also are found at this time. This syndrome is called desert rheumatism. Less than 1 percent of such cases go on to develop clinically disseminated disease, but, when they do, about one-third have severe destructive involvement of bone and joint. Chemotherapy and antibiotics (isoniazid, *p*-aminosalicylic acid, and streptomycin for tuberculosis, Amphotericin B for coccidioidomycosis) in combination with surgical removal of infected tissue is the treatment of choice.

Tendon sheaths and bursae become infected by the same organisms that cause arthritis and by similar mechanisms. They may or may not be accompanied by arthritis. Tenosynovitis and bursitis are considerably less frequent than the joint disease.

Arthralgias (pains in joints) and mild transient polyarthritis are frequent in the early phases of serum and infectious hepatitis. Evidence has been accumulating that at least some of the joint complaints result not so much from infection of the synovium as from deposition within it of immune complexes formed between the hepatitis B (Australia) antigen and antibody and complement. The observations have aroused speculation that this mechanism may be a prototype for other forms of rheumatic disease in which no specific agent has yet been recovered.

REFERENCES

ALPERT, E., ISSELBACHER, K. J., AND SCHUR, P. H. The pathogenesis of arthritis associated with viral hepatitis. Complement-component studies. N. Engl. J. Med. 285:185, 1971.

KEISER, H., RUBEN, F. L., WOLINSKY, E., AND KUSHNER, I. Clinical forms of gonococcal arthritis. N. Engl. J. Med. 279:234, 1968.

LONDON, W. T., FUCCILLO, D. A., ANDERSON, B., AND SEVER, J. L. Concentration of rubella virus antigen in chondrocytes of congenitally infected rabbits. Nature 226:172, 1970.

SOMERVILLE, E. W., AND WILKINSON, M. C. Girdlestone's Tuberculosis of Bone and Joint, 3rd ed. London, Oxford University Press, 1965.

WALDVOGEL, F. A., MEDOFF, G., AND SWARTZ, M. N. Osteomyelitis. Clinical Features, Therapeutic Considerations, and Unusual Aspects. Springfield, Ill., Charles C Thomas, 1971.

11

NON-INFECTIOUS ARTHRITIS

Rheumatic diseases rarely threaten life but, outside of mental illness, are our principal source of chronic disability. The National Health Survey in 1962 indicated that more than one million Americans were unemployable because of arthritis and rheumatism.

Some commonly used terms should be defined at the outset to minimize confusion. *Arthritis, tenosynovitis, tendinitis,* and *bursitis* are self-explanatory; they mean inflammation of a joint, tendon sheath, tendon, and bursa, respectively. *Synovitis* necessarily is part of arthritis but implies that the inflammation is confined to synovial tissue. *Arthralgias* are pains in joints and not necessarily associated with inflammation. *Rheumatoid arthritis* is a specific disease as is *rheumatic fever*, but *rheumatism* is another matter. It is a generic term for all manners of disturbance of the articular apparatus (joints, tendons, bursae, tendon sheaths), and, in addition, certain systemic diseases in which these structures are involved, or which have other clinical or anatomic similarities to the preceding, but little or no articular manifestations. Among the latter are some of the *"collagen diseases"* to be discussed subsequently. The domain of rheumatology varies in different countries. In the Americas and much of western Europe, it is a branch of internal medicine that deals with arthritis, collagen diseases, gout, and degenerative joint disease. Patients with osteoporosis and osteomalacia are cared for mostly by endocrinologists; polymyositis and muscular dystrophies by neurologists; metastatic bone tumors by oncologists, and surgical aspects of all these by orthopedists. This is not necessarily true elsewhere. The French rheumatologist, for example, also sees the metastatic bone problems while in Scandinavia patients with degeneraative joint disease are treated by the orthopedist.

RHEUMATOID ARTHRITIS

Several types of chronic arthritis resemble infectious joint disease pathologically and clinically but microorganisms are not recovered from them. In the absence of more reliable etiologic information, one distinguishes among them by somewhat imprecise features. Rheumatoid arthritis, the most important one, defies satisfactory definition. The American Rheumatism Association has established criteria according to which rheumatoid arthritis is diagnosed with increasing degrees of confidence: possible, probable, definite, and classical. Depending on the criteria employed, 1 to 5 percent of the population has rheumatoid arthritis. It is a chronic deforming disease, usually polyarticular and symmetrical. Any joint may be affected but the wrists and proximal parts of the fingers particularly so. The disease occurs principally in adults and more often in women than in men. The onset usually is insidious. There is little fever but systemic manifestations in the form of anemia and profound fatigue are common. A sensation of stiffness upon awakening in the morning is quite typical. The joints become swollen and painful while the inflammation is active. The end result is a variable degree of subluxation or ankylosis. Tendinitis and tenosynovitis make their own contributions to the deformities. The course may be either episodic or sustained and a small minority of patients enter into a complete remission. In most cases the patient must learn to live with the disease for the remainder of life. Over the long run, about 10 percent are incapacitated by it.

The earliest changes are found in the synovium and the inflammation evolves through a series of non-specific episodic and progressive sequences. Congestion and edema of the synovium are followed quickly by proliferation of granulation tissue and infiltration of chronic inflammatory cells (Fig. 11.1). Lymphocytes predominate initially but are replaced by plasma cells after a few months. Polymorphonuclear leukocytes are ordinarily few in the synovium but dominate the synovial fluid cell population during acute episodes. Synovial lining cells of both types become hyperplastic as well as enlarged. The enlarged A cells contain many lysosomal granules. As a result of the infiltration and growth of the granulation tissue, the synovium becomes thickened and thrown up in papillary folds. This pattern of hypertrophic villous synovitis, although characteristic, is not pathognomonic of rheumatoid arthritis. Extensive ulceration of the synovial surface occurs and necrotic masses of fibrin-rich synovial tissue are cast off as "rice bodies" into the joint fluid.

With the passage of time, synovial granulation tissue extends over and adheres to the surface of the cartilage and thus becomes a pannus (Fig. 11.2). Like hypertrophic villous synovitis, pannus formation is not specific morphologically and occurs at times in conditions other than rheumatoid arthritis. The pannus elaborates a collagenase which participates in the destruction of joint cartilage. Vascular granulation tissue also invades the cartilage from the subchondral bone marrow. The

116 *Musculoskeletal System*

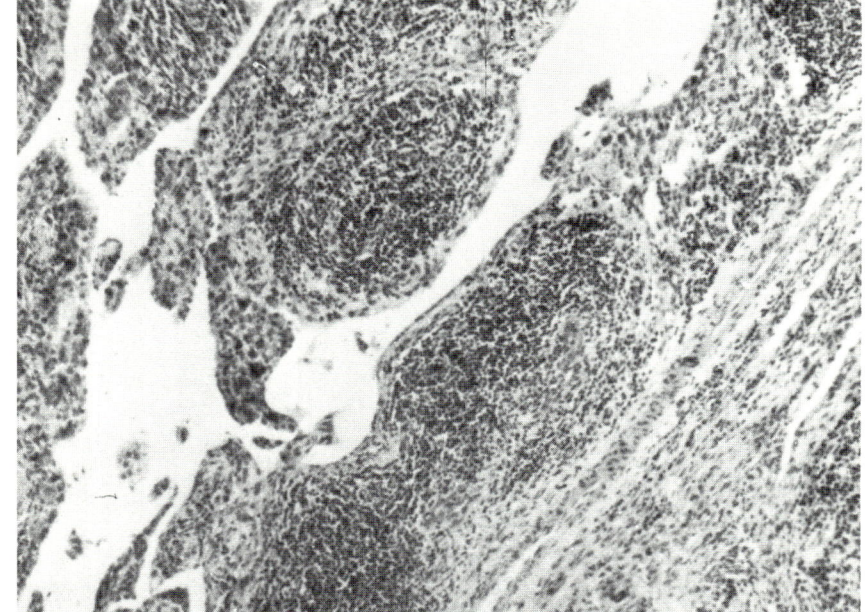

Figure 11.1. Hypertrophic villous synovitis, rheumatoid arthritis. The tissue is infiltrated with large numbers of chronic inflammatory cells disposed both in nodular and diffuse fashion. Enlarged lining cells form a distinct layer. (Hematoxylin and eosin, ×86.)

subchondral inflammation is responsible for the reduction of bone density that is seen early in rheumatoid arthritis next to the joint (Fig. 11.3). Clinically this is often loosely called para-articular osteoporosis. Later, destruction of the cartilage causes the joint space to become narrowed and the surface of the bone to be irregular. Weakening of the capsular and ligamentous supports allows the apposed joint surfaces to become partly dislocated from each other, that is, subluxated (Fig. 11.4). Fibrous adhesions between the ulcerated surfaces cause fusion (ankylosis) and immobility of the joint. Under some conditions, the fibrous tissue undergoes ossific metaplasia and bony ankylosis results (Fig. 11.5).

Rheumatoid Factors

In approximately 70 percent of patients with rheumatoid arthritis there occur one or more abnormal serum globulins called rheumatoid factors. Their distinguishing feature is a reactivity with IgG, and more particularly with the Fc portion of the heavy chain. They thus have features of autoantibodies. The most readily demonstrated rheumatoid

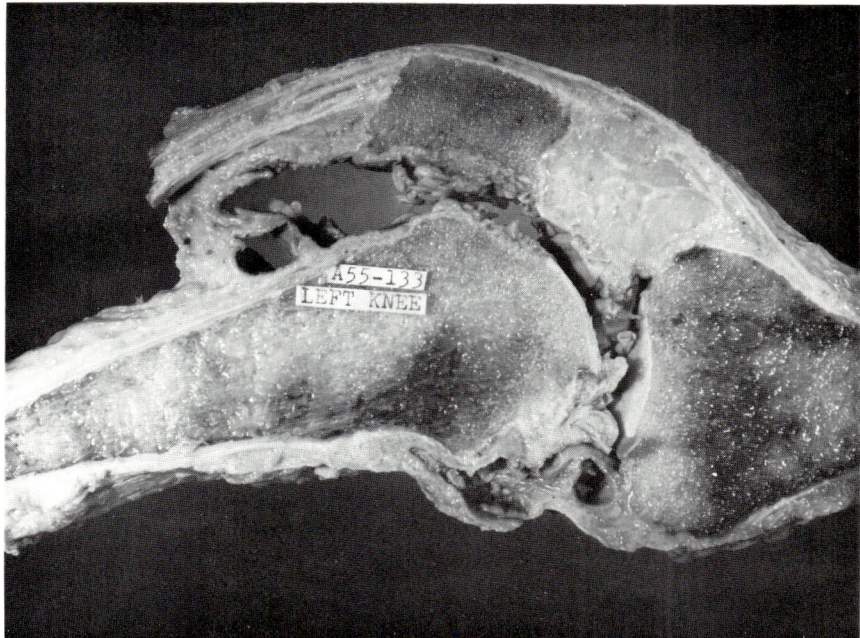

Figure 11.2. Rheumatoid arthritis of knee (sagittal). Thickened villous synovial tissue in the suprapatellar pouch is continuous with the pannus that replaces the articular cartilage of the patella and apposed facet of the femur.

factor is an IgM macroglobulin having a sedimentation coefficient variously of 19 or 22 S. Rheumatoid factor at times also is found in IgG and IgA fractions. These factors form the basis of widely employed diagnostic tests. In most test systems, an inert carrier particle, such as latex or the clay, bentonite, or sheep red blood cells, are coated with a modified IgG. In the presence of serum containing rheumatoid factor, these particles aggregate and become visible. A titer of 1:160 is positive in the latex fixation and above 1:32 in the bentonite flocculation test. These are not pathognomonic of rheumatoid arthritis and there may be false positive as well as false negative tests. Positive tests, even in high titer, are common in other inflammatory diseases associated with hypergammaglobulinemia, *e.g.*, trypanosomiasis, liver disease, and subacute bacterial endocarditis. Negative reactions occur in roughly 30 percent of rheumatoid patients. In general there is a correlation between the titer and the severity of the disease.

Rheumatoid factors are synthesized in the plasma cells of the synovial infiltrate. This is demonstrated by localization of fluorescein-labeled IgG over the cytoplasm of these cells. Complexes of rheumatoid factor and

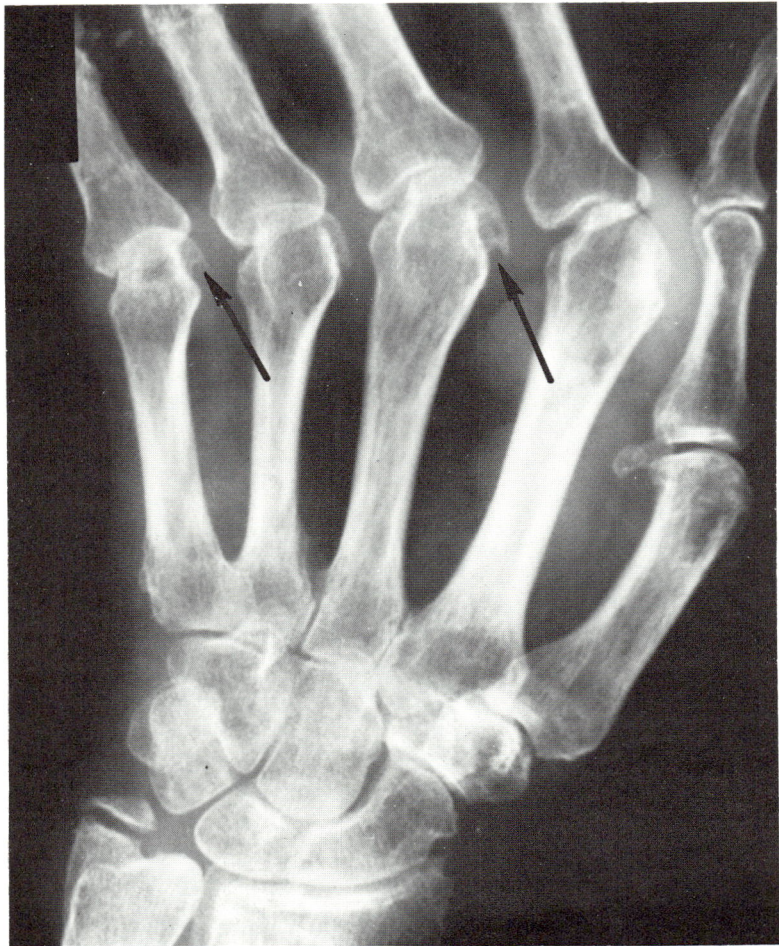

Figure 11.3. Rheumatoid arthritis of hand, radiologic grade 2 (mild to moderate). The important features are: (1) juxta-articular loss of bone density, (2) erosions of the joint surfaces (*arrows*), (3) narrowing of the metacarpophalangeal joint "spaces," and (4) ulnar deviation of the proximal phalanges.

IgG are phagocytosed by Type A lining cells and by polymorphonuclear leukocytes in synovial fluid. The latter have refractile cytoplasmic granules, actually phagolysosomes containing the complexes, and are called RA (rheumatoid arthritis) cells or ragocytes. Complement in fluid is reduced in seropositive cases while serum levels generally are normal. The synovial fluid contains increased numbers of cells, as many as 60,000 per mm^3 during periods of activity of the disease. Neutrophils are

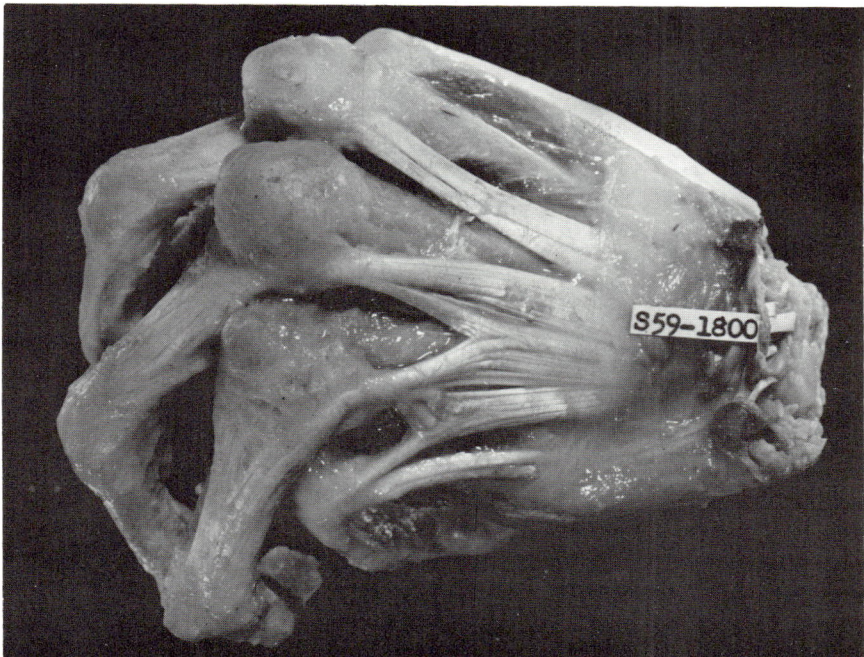

Figure 11.4. Characteristic hand deformities, rheumatoid arthritis. The proximal phalanges are subluxated beneath the metacarpal heads. The tendons are displaced from their normal positions over the center of the joints in association with ulnar deviation of the fingers.

numerous during acute episodes but mononuclear cells and lymphocytes may constitute the majority at other times. Viscosity is reduced and there is a poor mucin clot. Many lysosomal enzyme activities of the fluid are elevated.

From the preceding findings, two steps in the pathogenesis of rheumatoid arthritis can be reconstructed. (1) There is an immunologic process in which rheumatoid factors are synthesized by the plasma cell infiltrate in the synovium. (2) Phagocytosis of the rheumatoid factor-IgG complexes activates the complement system. Not only is complement consumed, lowering the levels in synovial fluid, but lysosomal enzymes are released and together with the complement cascade are instrumental in damaging the joint tissues. Two important modalities of drug therapy are directed at these steps: corticosteroids and immunosuppression against the former, and stabilization of lysosomal membranes by compounds such as gold and the antimalarials against the latter.

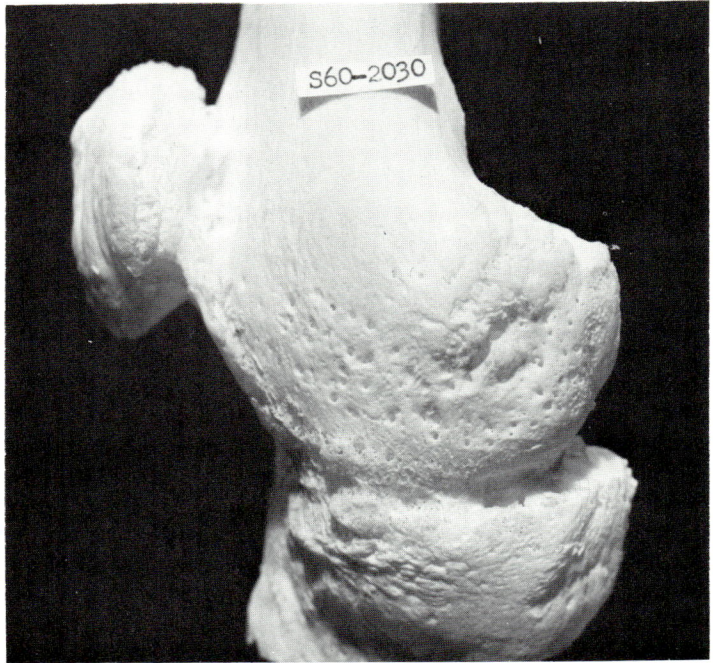

Figure 11.5. Bony ankylosis of knee, rheumatoid arthritis.

Extra-articular Lesions

Despite the systemic nature of rheumatoid arthritis, anatomic evidence of extra-articular involvement is relatively scanty. Subcutaneous nodules, ranging in diameter from a few millimeters to over 2 cm, occur over pressure points such as the olecranon process in about 20 percent of patients. They are found, as are other extra-articular lesions, almost exclusively in seropositive cases. The nodules are not tender, and, once formed, may persist indefinitely. Microscopically, as the name suggests, they resemble the subcutaneous nodules of rheumatic fever. Geographic areas of fibrinoid necrosis are present and they are surrounded by a palisade of elongated connective tissue cells (Fig. 11.6). There are differences between the subcutaneous nodules in the two disorders. Not only is the rheumatoid nodule larger, but it has more extensive necrosis and fibrosis. The process resembles that seen in the ulcerated, necrotic foci in synovial tissue. The difference is that the necrotic material has no joint space into which it can escape.

Pericarditis, as indicated by fibrous obliteration of the pericardial cavity, is found at necropsy in a quarter of rheumatoid patients but only infrequently does it cause clinical symptoms. Inflammation of small arteries occasionally leads to peripheral neuropathy or circulatory disturbances in the skin and splanchnic organs. At one time secondary amyloidosis was a common complication of rheumatoid arthritis, but it is not so often seen at present.

Still's Disease

Still's disease is the form of rheumatoid arthritis seen in childhood. It sometimes develops even in the first months of life. Juvenile rheumatoid arthritis resembles the adult disease in most respects but the usual serologic tests for rheumatoid factor are negative and there is a tendency for fewer joints to be involved. A skin rash and iridocyclitis are found with some frequency. Spinal involvement occurs so often that it seems likely that two disorders are being lumped under the term Still's disease: juvenile rheumatoid arthritis and juvenile ankylosing spondylitis. The recent finding of a high prevalence of a specific histocompatability antigen (HL-A W27) in lymphocytes of patients with Still's disease, as there also is in ankylosing spondylitis and anterior uveitis, lends support to this. Apart from such findings, there is little evidence for an inherited predisposition to rheumatoid arthritis.

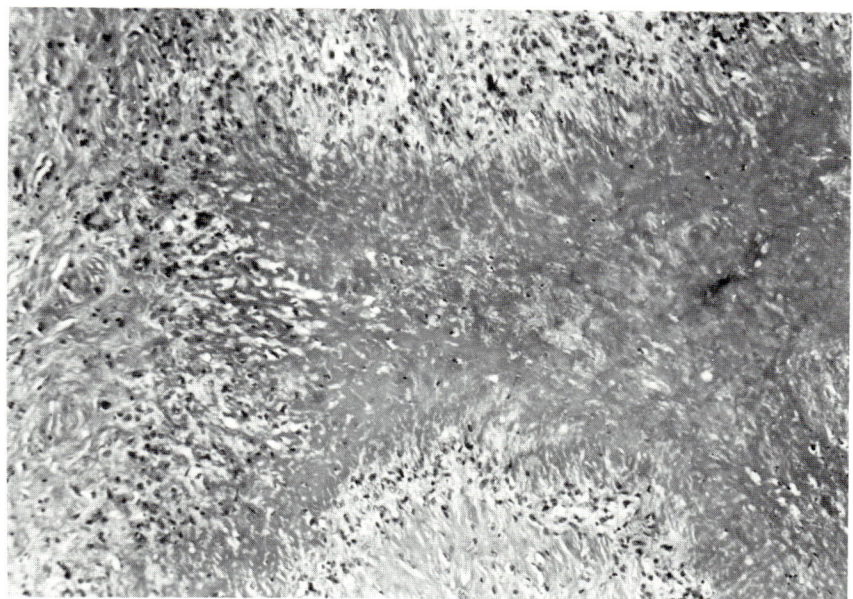

Figure 11.6. Subcutaneous nodule of rheumatoid arthritis. A palisade of elongated connective tissue cells is oriented at a right angle to the central zone of fibrinoid necrosis. (Hematoxylin and eosin, ×95.)

An enormous amount of research has failed to disclose the cause of rheumatoid arthritis. Current thinking on the subject may be divided into two theories: rheumatoid arthritis is an infection by agents that are difficult to culture; or an autoimmune disorder in which a presently unidentified event makes a specific immunoglobulin or joint component antigenic. There is no generalized abnormality of the immune system.

THE "VARIANTS" OF RHEUMATOID ARTHRITIS

Several forms of chronic deforming polyarthritis, once considered variant patterns of rheumatoid arthritis, now are generally classified as separate entities. All are seronegative and subcutaneous nodules practically never develop.

Psoriatic arthritis occurs in 5 percent of persons who have the distressing skin disease, psoriasis. The articular and cutaneous manifestations of psoriasis generally but not necessarily occur in synchrony and parallel each other in severity. The peripheral joint problems resemble those of rheumatoid arthritis clinically and pathologically but differ in two respects: (1) the distal rather than the proximal interphalangeal joints bear the brunt of the disease, and (2) resorption of bone near the joints is often more severe in psoriatic arthritis. It may progress to the point that the skin folds up over the destroyed bone and the tip of the finger can be pushed in or pulled out like an opera glass—the opera glass deformity. Constitutional symptoms like those of rheumatoid arthritis are present. Paraspinal ossification also is seen in some patients and may be confused with ankylosing spondylitis. The latter condition is the other major "variant" of rheumatoid arthritis and is discussed systematically with diseases of the spine in Chapter 13.

Peripheral and spinal joints are inflamed in some patients with chronic intestinal diseases, particularly ulcerative colitis and less often regional enteritis and Whipple's disease. The arthritis is less severe than rheumatoid arthritis but at times residual loss of mobility or deformity results. The histologic changes are like those of rheumatoid arthritis but milder; both are non-specific. The microorganisms seen in submucosal mononuclear cells in Whipple's disease have not been found in the synovium.

Chronic arthritis curiously is also found in approximately one-third of patients with congenital or acquired agammaglobulinemias and hypogammaglobulinemias. Most cases are congenital and all classes of immunoglobulin are affected. When the immunologic defect is confined to one type, it is the IgA deficiency that is most likely to be associated with the rheumatic syndrome. The mechanism of the joint involvement is unknown. Perhaps the absence of antibody formation permits joints to become infected, but the fact is that microorganisms have not been recovered. The clinical picture closely resembles Still's disease. The morphologic appearance of the synovium also is like that of rheumatoid arthritis except for the absence of plasma cells necessary for the production of immunoglobulins (Fig. 11.7). Immunologic theories of rheumatoid arthritis must reckon with this phenomenon.

THE COLLAGEN DISEASES

The scope and definition of "collagen diseases" also are somewhat vague. Inherent in them is the idea of lesions widely disseminated in collagenous connective tissues. As originally conceived, these lesions involve fibrinoid change, that is, the fibers come to resemble fibrin but are not necessarily fibrinous. Now it is known that not all fibrinoid materials have the same composition or pathologic significance. Despite this, there does exist considerable overlap among these diseases and clinically they are frequently distinguished only with difficulty. Fibrinoid necrosis has already been described as a feature of the subcutane-

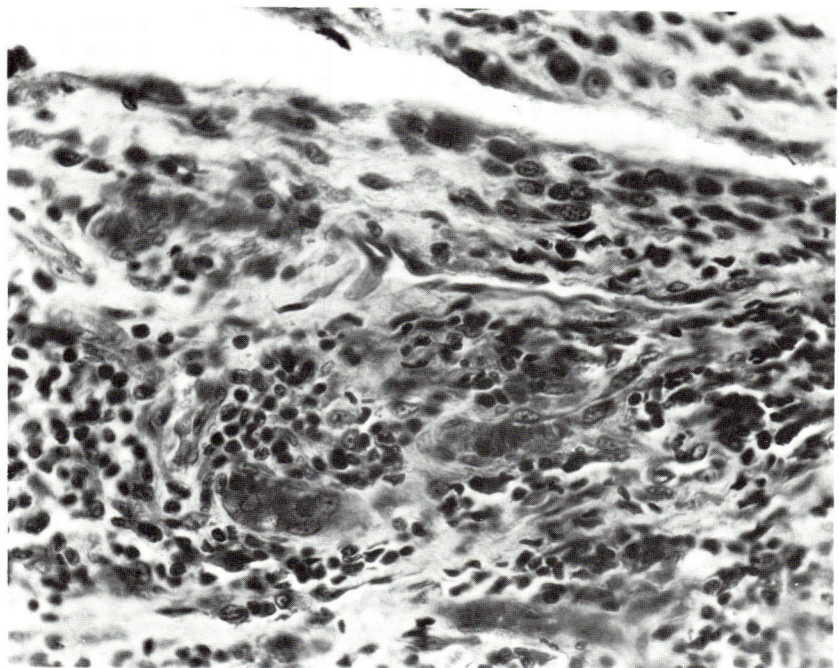

Figure 11.7. Chronic synovitis in congenital hypogammaglobulinemia. There is a proliferation of large Type B lining cells as well as fibroblasts and capillaries. The inflammatory infiltrate contains mainly lymphocytes but no plasma cells. (Hematoxylin and erosin, approximately ×350.)

ous nodules of rheumatoid arthritis and rheumatic fever, and these conditions are sometimes classified among the collagen diseases. The entities universally included are systemic lupus erythematosus, generalized scleroderma, dermatomyositis, and polyarteritis nodosa. There is no reason to believe that the collagen itself is abnormal in any of them and the name "connective tissue disease" or "collagen-vascular disease" is preferred by some because of this. The manifestations are protean and they are dealt with under multiple organ systems. Musculoskeletal complaints occur with different frequencies in each and it is to these that the following descriptions are directed.

Systemic Lupus Erythematosus (SLE)

SLE is the most frequent of the collagen diseases. It occurs primarily in young to middle-aged women and roughly one-tenth die of it within 10 years. Originally recognized and named for a characteristic skin eruption (butterfly rash over the malar region), the principal lethal manifestations are lupus nephritis and cerebral vasculitis. A spectrum of serologic abnormalities indicates that a generalized disturbance of the immunologic mechanism is at fault. Several antibodies to DNA and its histone protein provide the clue to the nature of the disease. The lesions of SLE result from the deposition of DNA-antiDNA complexes in the tissues and the activation there of the complement system. Focal fibrinoid change and extracellular DNA (hematoxylin bodies) are seen microscopically at these sites in post-mortem specimens. The antibodies to DNA protein form the basis of the lupus erythematosus (LE) cell test: when a damaged cell, usually a polymorphonuclear leukocyte, is exposed *in vitro* to serum containing such an antibody (IgG), the nucleus swells and becomes homogenized. In turn, the altered nucleus is phagocytosed by another neutrophil and appears as an inclusion body within its cytoplasm (Fig. 11.8). LE cells do not form in living tissues so they must be regarded as a useful diagnostic artifact. They are found in the vast majority of cases of SLE (90 percent) but are not pathognomonic. Other and simpler immunologic methods for demonstrating antinuclear antibodies are now widely used. They are virtually always positive in SLE but are often positive in other diseases as well. Thus these antinuclear antibody tests are good screening procedures and a negative test puts the diagnosis of SLE in doubt. Serum complement levels are low. Other evidences of SLE being an autoimmune disease include antibodies against several classes of RNA, Coombs-positive anemia, and false positive serologic tests for syphilis. Drugs such as the antihypertensive compound, hydralazine, and the anticonvulsant, dilantin, induce a syndrome closely resembling SLE. SLE-like illness sometimes occurs in patients with chronic active hepatitis. LE cells are present, but the

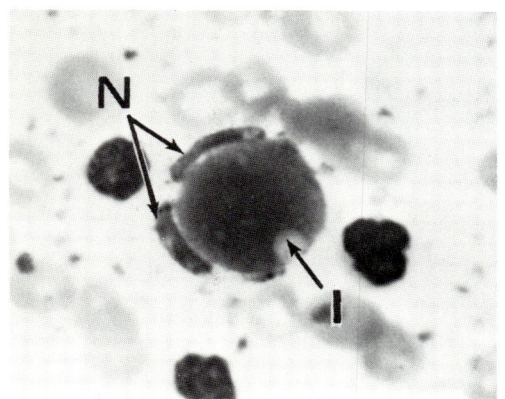

Figure 11.8. Lupus erythematosus (LE) cell. A rounded, homogeneous inclusion body (*I*) is engulfed by a bilobed polymorphonuclear neutrophil (*N*). (Wright stain, ×1200.)

hepatitis B antigen is not. Another possible etiologic factor being studied is a cytoplasmic tubular inclusion seen with the electron microscope in glomerular endothelial cells and circulating lymphocytes. They represent evidence of past viral infection but are not themselves viruses. Cytotubular bodies are sometimes seen in other diseases.

Arthralgias or mild polyarthritis occur in 75 percent of patients with SLE during its initial phase. In cases that go on for many years, articular changes often become progressive and resemble low grade rheumatoid arthritis. The relationship to rheumatoid arthritis is made more complex by the fact that one-fourth of patients with SLE have positive tests for rheumatoid factor and as many rheumatoid patients have positive LE cell tests. Nevertheless the clinical course of the two diseases usually is distinct and not influenced by the overlap in the serologic abnormalities.

Scleroderma

Like SLE, generalized scleroderma is a complicated systemic disease which has prominent cutaneous manifestations. There are approximately 20,000 cases in this country. The skin of the face and hands becomes tightened and rigid because of fibrous thickening of the dermis. Fibrosis of the gastrointestinal tract, particularly of the esophagus, and of the myocardium also occur. For this reason the name "progressive systemic sclerosis" is often advocated. It usually appears during early middle-age and has a slowly progressive course. At times it terminates in fulminant manner with malignant hypertension. Raynaud's phenomenon, a paroxysmal spasm of the digital arteries precipitated by exposure to cold, is another vascular abnormality and it is found in 90 percent of cases. Presumably, there is some pathogenetic relationship between the circulatory changes and the fibrosis.

Most of the restriction of joint motion in scleroderma is attributable to the tightening of the skin of the fingers (acrosclerosis). There also frequently occurs a true fibrinous synovitis of the fingers, wrists, knees, and ankles. The joints become swollen and stiff but destructive changes are uncommon. There are no distinctive immunologic abnormalities in scleroderma. The majority of patients do have, however, positive antinuclear antibody tests and one-third have positive rheumatoid factor tests in low titer.

Polyarteritis Nodosa (PAN)

This is an idiopathic systemic disease characterized by segmental fibrinoid necrosis and inflammation of medium sized (1- to 2-mm external diameter) arteries. All parts of the body may be affected, but the kidneys, myocardium, and peripheral nerves particularly so. The manifestations are highly variable depending on the site involved but the prognosis is uniformly poor. Most patients die within 1 year.

Muscle soreness and arthralgias are common but swelling or other objective changes in joints are not. The principal interest in PAN for the rheumatologist is that somewhat similar lesions are frequent in the other collagen diseases, and also in rheumatoid arthritis and rheumatic fever. There is less resemblance to the arteritis of polymyalgia rheumatica. The arteries involved in rheumatoid arthritis and the collagen diseases usually are of smaller caliber (0.1- to 0.2-mm diameter) and the histologic changes less explosive. Disseminated arteritis is thus an additional component of the collagen ("collagen-vascular") diseases. It may be that PAN is not a single entity but an extreme expression of different types of arteritis including those associated with the rheumatic diseases. The peripheral neuropathy sometimes seen in rheumatoid arthritis, for example, is secondary to segmental arteritis in the nerve. It resembles the neuropathy of PAN clinically and pathologically.

Sjögren's Syndrome

This syndrome occurs principally in older women and comprises an unlikely triad of abnormalities: (1) keratoconjunctivitis sicca: dryness of the eyes related to diminished lacrimation; (2) xerostomia: dryness of the mouth resulting from diminished salivary gland secretion; and (3) rheumatoid arthritis or one of the collagen diseases. The relationship between the glandular and rheumatic changes seems quite remote but multiple serologic abnormalities, analogous to those of SLE, occur in Sjögren's syndrome. The secretory difficulties result from distortion and occlusion of intralobular ducts by lymphocytic infiltration and epimyoepithelial metaplasia of the lining cells. As many as 90 percent of patients have positive serologic tests for rheumatoid arthritis even when there is no joint disease. Sixty-eight percent have antinuclear antibodies in the serum. Antibodies to a cytoplasmic antigen in the salivary duct epithelium also are frequent. These and other autoantibodies are produced by the lymphocytic infiltrates in the glands.

Rheumatic Fever

There has been a dramatic reduction in the prevalence of acute rheumatic fever over the past 3 decades but rheumatic valvular heart disease still is quite frequent. Occult carditis therefore exists and progresses in the absence of articular or other clinically obtrusive manifestations. The cause of rheumatic fever is unknown except for the fact that it follows shortly on Group A β-hemolytic streptococcal pharyngitis. Serum titers of several antibodies directed against streptococcal products, e.g., antistreptolysin O, are elevated in incipient rheumatic fever. They are of diagnostic value but their role in the pathogenesis of the disease has not been established. Subcutaneous nodules and polyarthritis accompany the fever but are transient and do not leave anatomic residua, unlike rheumatoid arthritis. The serologic tests for rheumatoid factor are negative and there is no etiologic relationship between these two disorders.

REFERENCES

ANDERSON, L. G., TARPLEY, T. M., TALAL, N., CUMMINGS, N. A., WOLF, R. O., AND SCHALL, G. L. Cellular-versus-humoral autoimmune responses to salivary gland in Sjögren's syndrome. Clin. Exp. Immunol. 13:335, 1973.

BLAND, J. H., AND PHILLIPS, C. A. Etiology and pathogenesis of rheumatoid arthritis and related multisystem diseases. Semin. Arth. Rheum. 1:339, 1972.

DUBOIS, E. L. (Ed.). Lupus Erythematosus, 2nd ed. Los Angeles, University of Southern California Press. 1974.

GARDNER, D. L. The Pathology of Rheumatoid Arthritis. Baltimore, Williams & Wilkins, 1972.

MARKOWITZ, M., AND GOLDIS, L. Rheumatic Fever. Diagnosis, Management and Prevention. 2nd ed. Philadelphia, W. B. Saunders, 1972.

RACHELEFSKY, G. S., TERASAKI, P. I., KATZ, R., AND STIEHM, E. R. Increased prevalence of W27 in juvenile rheumatoid arthritis. N. Engl. J. Med. 290:892, 1974.

RODNAN, G. P., MCEWEN, C., AND WALLACE, S. L. (Eds.) Primer on the rheumatic diseases, 7th ed. J. A. M. A. 224 (Suppl.) 661, 1973.

SHEARN, M. A. Sjögren's Syndrome. Philadelphia, W. B. Saunders, 1971.

SOKOLOFF, L. The pathophysiology of peripheral blood vessels in collagen diseases. In The Peripheral Blood Vessels. Int. Acad. Pathol. Monogr. 4:297, 1963.

ZIFF, M. Pathophysiology of rheumatoid arthritis. Fed. Proc. 32:131, 1973.

12

NON-INFLAMMATORY JOINT DISEASES

A large number of biologic and mechanical, as distinct from inflammatory, abnormalities also cause disabling joint disease. Unless corrected, they terminate in osteoarthritic degeneration.

OSTEOARTHRITIS

Despite the name, this exceedingly frequent disorder is not inflammatory and many now prefer to call it degenerative joint disease or arthrosis. It affects virtually everyone by the time of middle age and becomes progressively more severe as the years go by. Although it does not cause major clinical complaints in most persons, the great prevalence of the lesions makes osteoarthritis the principal cause of rheumatic disability in the gainfully employed as well as the aged population. Two pathologic processes must be distinguished in its development: mechanical disruption of the bearing surface of the articular cartilage, and proliferation of new bone beneath and at the margin of the latter. At an early stage, the cartilage splits along its collagenous arcades so that the surface appears fibrillated rather than smooth. Progressive abrasion of the disrupted cartilage exposes the subchondral bony plate, and the latter becomes polished. This is associated with reactive proliferation of bone in the adjacent marrow. Together these processes cause the surface to look like ivory, *i.e.*, be eburnated. New bone also is formed at the margin of the cartilage where there normally is an intimate blending of vascular synovium, capsule, periosteum, and bone. The outgrowth of bone, the osteophyte, actually is capped by newly formed cartilage, another example of the pluripotentiality of nascent skeletal connective tissue. In advanced cases, marked deformity occurs as the subchondral bone undergoes remodeling, microfractures, and pseudocystic degeneration (Fig. 12.1). Ankylosis is not a feature.

A sizeable proportion of cases are of secondary type, *i.e.*, they develop as a consequence of some pre-existing abnormality of the joint: structural

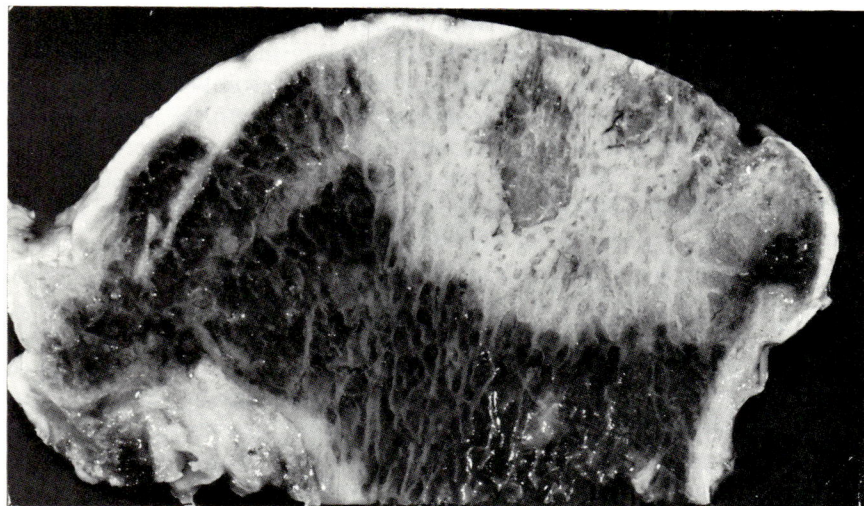

Figure 12.1. Advanced osteoarthritis, head of femur. The normal globoid contour has been deformed both by loss of white articular cartilage from most of the joint surface and by formation of marginal osteophytes. The spur at the left has grown not only to the side but superiorly into the residual original cartilage. Subchondral pseudocysts approach the eroded joint surface through slender crevices. The pallor of the eburnated tissue reflects the condensation of bony trabeculae and fibrous tissue in contrast to the darker, vascular hematopoietic marrow.

malformations, trauma, or arthritis. The majority of cases, however, are idiopathic and no antecedent cause can be identified. In these instances, some biologic aging changes in the joint tissues or cumulative microtraumata may be the important etiologic factors. The nature of the aging change is not known. The inability of chondrocytes to repair cartilage is the cornerstone of the "wear-and-tear" concept of degenerative joint disease. With the exception of senescent pigmentation, no changes in the composition or physical properties of intact adult articular cartilage have been found with aging. In fibrillated cartilage, the quantity of proteoglycan relative to the amount of collagen is reduced. The synovial fluid is normal unless osteoarthritis is complicated by synovitis. It often contains minute fragments of cartilaginous debris.

Much of the pain in osteoarthritis results from circulatory engorgement of the remodeling subchondral bone or from small tears that develop in the capsule or ligaments. A foreign body reaction to shards of cartilage displaced into the synovium accounts for the effusion seen in some cases as well as for pain. The osteophytes create a mechanical impediment to motion but also enlarge the contact area supporting the urine. Under alkaline conditions, the excreted homogentisic acid is oxidized and polymerizes into a melanin-like pigment; hence the alkaptonuria. After many years, hyaline cartilages within the body undergo similar pigmentation. The ear and tracheobronchial cartilages appear dark but the principal harmful consequences of the defect are found in the articular system. The physical properties of the ochronotic cartilage change; it becomes very brittle and as a result there occurs a severe and generalized osteoarthritic degeneration of peripheral joints and intervertebral discs. As much as 6 inches of height may be lost as the discs collapse. The lesion, despite its infrequency, is of interest as a prototype of biologic mechanisms that may influence the development of osteoarthritis.

Hemophilic arthropathy represents, aside from life-threatening hemorrhages, the principal long-term medical problem of the hemophilias. It occurs in approximately 90 per cent of affected persons of whom there are approximately 25,000 in the United States. Bleeding takes place, often following little trauma, into the synovial cavity and also into subchondral bone and periarticular tissues. It causes much pain and there may be considerable local inflammation. This may be confused with Still's disease in children. Although the acute episodes usually subside within a few weeks, repeated hemarthroses result in degenerative changes resembling osteoarthritis. The synovial tissue contains large amounts of hemosiderin but the explanation for the deterioration of the cartilage is not obvious. In particularly severe cases, destruction of the joint passes beyond what one would expect in osteoarthritis (Fig. 12.3). Because of pain and immobilization, marked disuse osteoporosis develops in the involved limb. Hemorrhage into bone and surrounding tissues sometimes leads to marked tumor-like proliferation as well as destruction of bone. This hemophilic pseudotumor is not to be confused with true neoplasms. The availability of clotting factors now opens new vistas for the prevention and orthopedic management of the hemophilias.

Neuropathic arthropathies (*Charcot joints*) are the consequence of defective articular proprioception. Syphilitic tabes dorsalis, diabetic neuropathy, and syringomyelia are the most frequent underlying neurologic causes. The afferent signals which normally tell muscles to stop a joint from moving beyond the limits of its strength are not there to prevent tearing of the capsular and ligamentous supports. This leads to a greatly exaggerated osteoarthritis, featured by microfractures and disintegration of the articular surface, and florid osteophyte formation (Fig. 12.4). The joint detritus lodges in the synovium and evokes a secondary synovial osteochondromatosis. Pain is relatively mild considering the extent of joint destruction and the principal manifestation is loss of

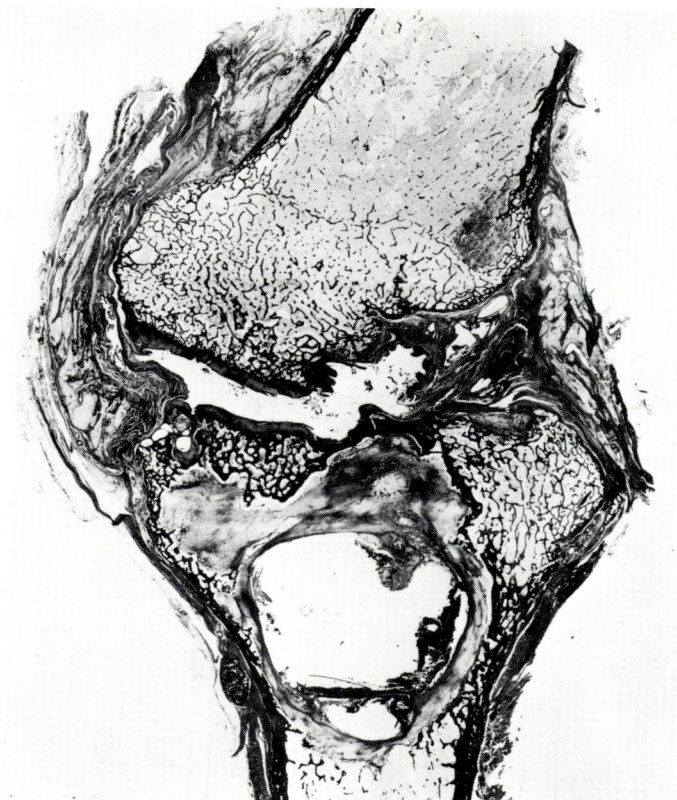

Figure 12.3. Advanced hemophilic arthropathy, sagittal section of knee. The femoral condyle appears flattened and the patella has disappeared. There is a large area of cystic resorption of the head of the tibia. The articular cartilage of most of the joint surface has been replaced by a synovial-like fibrous tissue. The synovial and capsular tissues are thickened and siderotic. Fibrous ankylosis is seen at the right. Note the disuse osteoporosis of the bone cortices. (Hematoxylin and eosin, $\times \sqrt[3]{4}$.)

mechanical stability. One or at most a few joints are involved and the location varies with the particular neural lesion. Usually it is the large joints of the lower extremity that are affected, but the dorsolumbar spine and feet may also be affected. The upper extremities are spared except in syringomyelia where there is a predilection for the elbow joint.

CRYSTAL DEPOSITION ARTHROPATHIES

Gout

Gout is not one but a group of disorders in which sodium acid (monosodium) urate crystals are deposited in and around joints and, to a

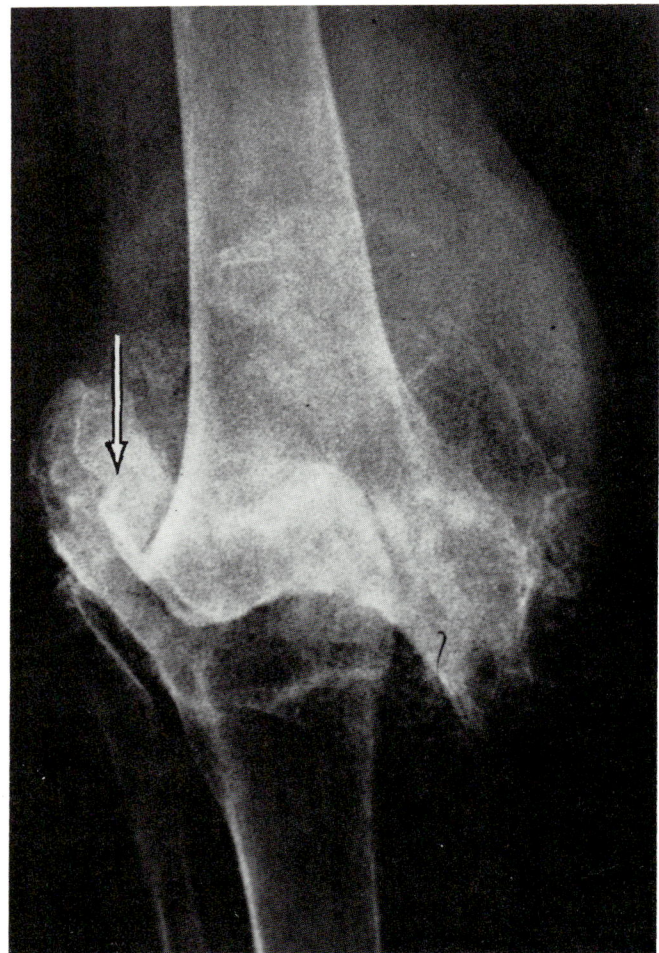

Figure 12.4. Severe neuropathic joint disease of 20 years' duration in tabes dorsalis. The condyles have lost much of their substance and the distal femur overrides the tibia. The shelf of bone projecting from the lateral margin of the tibial plateau is a massive marginal osteophyte *(arrow)*. The mottled densities in the adjacent soft tissues are shards of fragmented bone and secondary synovial osteochondromata in the enlarged joint lining.

lesser extent, other tissues (Fig. 12.5). The excessive quantities of urate arise through several mechanisms: metabolic overproduction (the *de novo* pathway), insufficient excretion, or presentation of a large pool of nucleic acid for degradation by some unrelated pathologic process (Fig. 12.6). Primary gout is the common variety in which the underlying

Non-inflammatory Joint Diseases   135

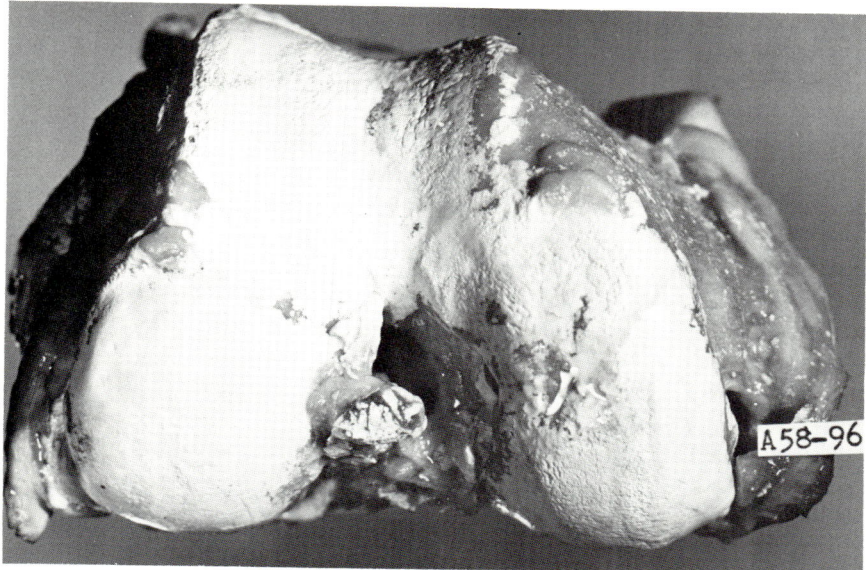

Figure 12.5. Chronic gouty arthritis, distal femur. White, chalk-like deposits of urate have been laid down on the surface of the cartilage and, to a lesser degree, the cruciate ligament and synovium.

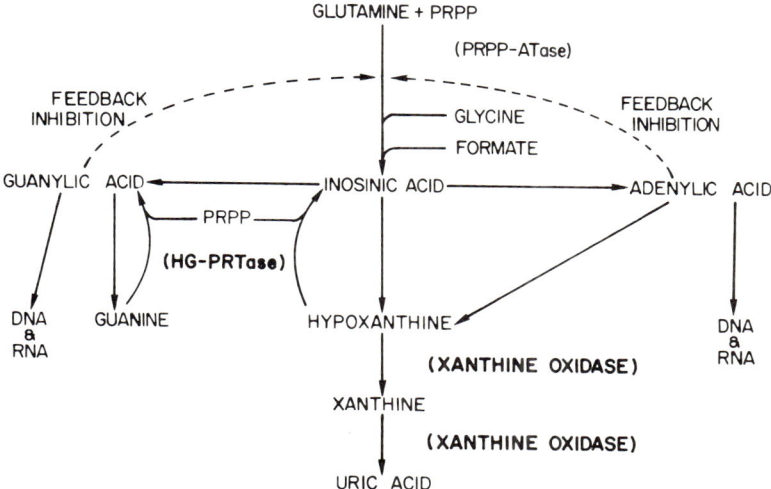

Figure 12.6. Biosynthetic pathways of uric acid. The enzymes in bold face are the two most pertinent to the clinical management of gout. PRPP, phosphoribosylpyrophosphate. PRPP-ATase, phosphoribosylpyrophosphate-amidotransferase. HG-PRTase, hypoxanthine-guanine-phosphoribosyltransferase.

hyperuricemia is the result of an inborn error of metabolism. The principal causes of secondary gout are myeloproliferative diseases such as leukemias, polycythemia vera, and agnogenic myeloid metaplasia, in which large numbers of nucleated blood cells undergo destruction. The serum urate level is influenced to a relatively small degree by the dietary intake of nucleic acids.

Hyperuricemia and gout are by no means the same thing, and only 10 percent of patients with established hyperuricemia have gout. Just what constitutes hyperuricemia is controversial. Part of the uncertainties in this area arise from variations in the techniques employed for measuring urate. Modern colorimetric and enzymatic methods give upper normal limits for men between 6.9 and 7.6 mg per 100 ml serum; automatic analyzers yield values 0.4 mg or more higher. The levels in women are lower than those in men. There is more to gout than the degree of hyperuricemia. Even in normal men, body fluids are close to supersaturated with urate. Precipitation of monosodium urate crystals, which is the essence of gout, requires additional mechanisms.

Overproduction of urate, as shown by excessive quantities of urate in the urine, occurs in about one-fourth of primary cases. The abnormal synthetic mechanisms involved are only partly understood. Urate is formed normally from hypoxanthine, the ultimate breakdown product of the nucleotides, inosinic, guanylic, and adenylic acids (Fig. 12.6). The first step in the synthesis of urate is the reaction between phosphoribosylpyrophosphate (PRPP) and glutamine to form a precursor of inosinic acid. This step is regulated through a feedback inhibition by the ribonucleotides. In the catabolism of the purines, two enzymes are of particular clinical importance. Hypoxanthine-guanine-phosphoribosyltransferase (HG-PRTase) catalyzes the reconversion of the purines, hypoxanthine and guanine, to their ribonucleotide forms. In the absence of HG-PRTase, there is no salvage of hypoxanthine or guanine and these bases are promptly converted to urate. HG-PRTase activity is readily measured in red blood cells and in cultured fibroblasts. Defective function of HG-PRTase and consequent hyperuricemia are demonstrated dramatically in the Lesch-Nyhan syndrome. This is a rare X-linked inherited deficiency of HG-PRTase and is manifested clinically by extreme hyperuricemia and bizarre behavioral disturbances: compulsive aggressiveness and self-mutilation. Most boys die of this when they are only a few years old but, if they live long enough, gouty tophi develop. A partial HG-PRTase deficiency is found in some adult cases of primary gout also.

The other important enzyme is xanthine oxidase which catalyzes the oxidation of hypoxanthine to xanthine, and xanthine to urate. A class of xanthine oxidase inhibitors (*e.g.*, allopurinol) is widely employed to reduce the amount of uric acid synthesized and excreted in gout. Unlike most mammalian species, the human has no enzyme for breaking the urate down further. Intestinal bacteria make a uricase but this does not enter the body.

Approximately two-thirds of the urate synthesized each day is excreted in the urine (Fig. 12.7). Most of the remainder enters the intestine where it is destroyed by the bacterial uricase. A reduced renal efficiency in excreting urate, alone or in conjunction with overproduction, is the mechanism responsible for approximately three-fourths of primary gout. There are three steps in the renal elimination of urate. Almost all of the plasma urate is first filtered through the glomerulus. It is then resorbed by the renal tubules. The third step is secretion of urate further along in the tubule, the same region involved in the clearance of lactic and keto (acetoacetic and β-hydroxybutyric) acids. It is thus not surprising that, in systemic acidosis, competition of these acids reduces the renal excretion of urate. This is the explanation for the triggering of acute gout by the starvation treatment of obesity or thiazide diuretics. Uricosuric agents that decrease the tubular resorption step—salicylates, probenecid, and sulfinpyrazone—are employed to increase the excretion of urate in chronic gout.

Gout is a relatively common disease. Its prevalence varies in different populations but it accounts for approximately 5 percent of patients seen

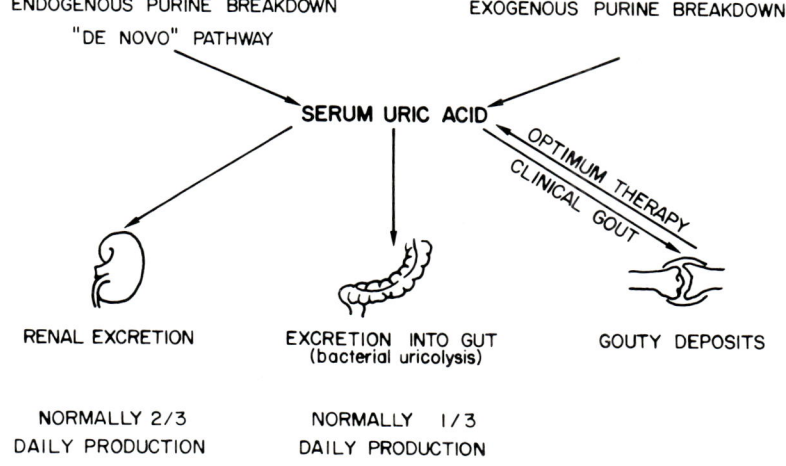

Figure 12.7. Avenues by which urates move in relation to gout: under-excretors by inefficiency of the renal mechanism. The intestinal route is a one-way street and contributes to gout only in disturbances of intestinal flora.

in arthritis clinics here. It is overwhelmingly a disease of men and primary types are not often seen in women. The inheritance is multifactorial and a familial occurrence can be elicited by careful questioning in the majority of cases. Gout has acute and chronic manifestations. The initial acute episode usually appears explosively in men with previously asymptomatic hyperuricemia after the age of 30. Typically a single joint is involved at first. The first metatarsophalangeal joint is the site of predilection and acute gouty arthritis here has been known as podagra for many centuries. Excruciating pain, swelling, and tenderness develop over the course of a few hours. The acute episodes untreated last from several days to several weeks. Early in the course of the disease, they go into complete remission for some years. Subsequently acute attacks occur at more frequent intervals and involve more and more joints. The serum urate concentration does not fluctuate with the clinical expression of acute gout, and local factors are involved in the precipitation of the crystals. The crystals are phagocytosed by polymorphonuclear leukocytes (Fig. 12.8B). Lysosomal enzymes are released and the Hageman factor, the plasma protein that initiates the intrinsic clotting mechanism, is activated. This in turn causes production of kinin peptides that mediate the vascular and other components of the acute inflammatory reaction. Colchicine has been used for ages in the treatment of acute gout. It does not affect the metabolism or excretion of urate but it is highly effective in relieving the acute inflammation. Colchicine and its analogues exert their effect by interfering with the polymerization of microtubular proteins of cells. The microtubules of leukocytes are instrumental in the phagocytic process and it is the suppression of the latter that interrupts the inflammatory sequence.

In chronic gouty arthritis, chalk-like deposits of monosodium urate accumulate with or without acute attacks. The crystals are acicular (needle shaped) and evoke a foreign body rather than a purulent inflammatory reaction. Multinucleated as well as mononuclear cells surround the crystals and in time they are encased in a hyalinized fibrous tissue. Such a deposit is a tophus. It occurs not only in synovial and periarticular tissues but also next to the cartilage in the helix of the ear. Tophi may become very large. In joints they cause a lesion very much like that of osteoarthritis except that the crystals are seen in the surface of the cartilage. In more advanced forms, large urate masses may obliterate the landmarks of the joint. Subchondral deposits destroy the bone. In x-ray films the resulting "punched out" lesions resemble the pseudocysts of osteoarthritis but lack their peripheral sclerotic rim. Progression of tophaceous gout can be retarded and even reversed by the continued use of the uricosuric agents and xanthine oxidase inhibitors.

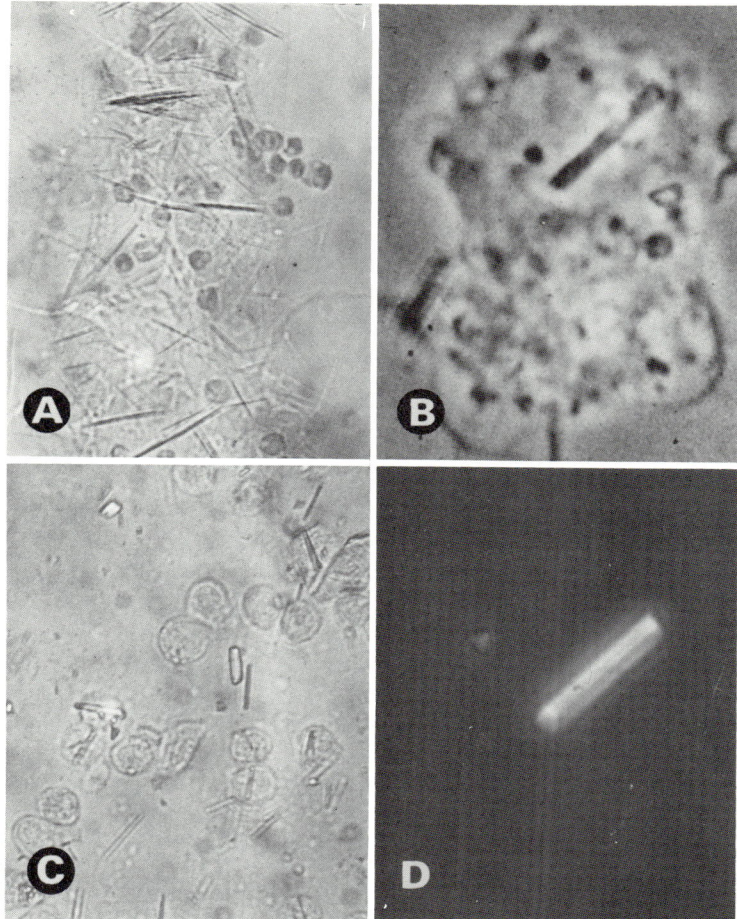

Figure 12.8. Crystals in synovial fluid. (*A* to *C*, phase contrast; *D*, polarized light.) *A*. Acicular urate crystals in gout (approximately × 400). *B*. Urate crystal within polymorphonuclear leukocyte in acute gout (approximately × 1500). *C*. Calcium pyrophosphate in pseudogout. The crystals are less needle-like than the monosodium urate (approximately × 400). *D*. Calcium pyrophosphate (approximately × 1000).

There are numerous systemic dysfunctions and ramifications of gout. Uric acid stones are frequent in the lower urinary tract. Smaller tophaceous deposits in the renal pyramids cause some cystic dilatation of the tubules as well as fibrosis (gout nephrosis). This favors the development of chronic pyelonephritis. For some reason, hypertension, hypertriglyceridemia, diabetes mellitus, and atherosclerosis often complicate chronic tophaceous gout. In chronic uremia with acidosis due to glomerulonephritis and other renal diseases, hyperuricemia is present for long periods of time, but secondary gout rarely supervenes. Hyperuricemia and even precipitation of amorphous urates in the collecting tubules are often seen in acute leukemias but they rarely lead to secondary gout. Obviously there is much yet to be learned about the pathogenesis of gout.

The synovial fluid contains urate crystals both during acute episodes and the tophaceous stage; in acute gout, many crystals lie within polymorphonuclear leukocytes. Identification of the crystals is an invaluable diagnostic procedure (Fig. 12.8A, B). The urates are needle-shaped and display a strong negative birefringence when examined with plane polarized light. This property is important in differentiating urate crystals from those of chondrocalcinosis.

Chondrocalcinosis

Chondrocalcinosis is characterized by deposition of calcium pyrophosphate crystals in joint cartilages (Fig. 12.9). Minute quantities of this or similar minerals are frequently observed at necropsy, particularly in the menisci of the knee. Usually they are confined to the cartilage and cause no discomfort. Larger deposits erupt into the joint space and evoke a gout-like syndrome (pseudogout). Although pseudogout has only recently come to the attention of rheumatologists, it is now seen at least as often as gout in arthritis clinics. This is a disease of the elderly. The sex differences of gout are not seen but there is an hereditary pattern in some cases. In acute episodes, the pyrophosphate crystals are found in the synovial fluid (Fig. 12.8C and D), and are distinguished from urates by being broader, more blunt, and by a weakly positive sign of their birefringence. The acute attacks do not usually respond to colchicine and remit spontaneously in a few weeks. In longstanding cases, osteoarthritic degeneration is superimposed. In x-ray films, radiopaque material is seen in the middle zone of the articular cartilage (Fig. 12.9). The nature of the metabolic abnormality awaits clarification. Pyrophosphate is a normal component of calcifying tissues and one or another local disturbance of alkaline phosphatase or pyrophosphatase may be at fault.

The type of mineral in other dystrophic calcifications—*calcific tendinitis, calcinosis cutis*, and *tumoral calcinosis*—varies from the preceding. In some it is hydroxyapatite, but in others calcium phosphate or calcium carbonate.

ASEPTIC NECROSIS OF BONE

Because of their extensive collateral circulation, epiphyses are not readily infarcted by occlusion of a single artery. Infarcts (aseptic necrosis), however, are quite common, particularly in the head of the

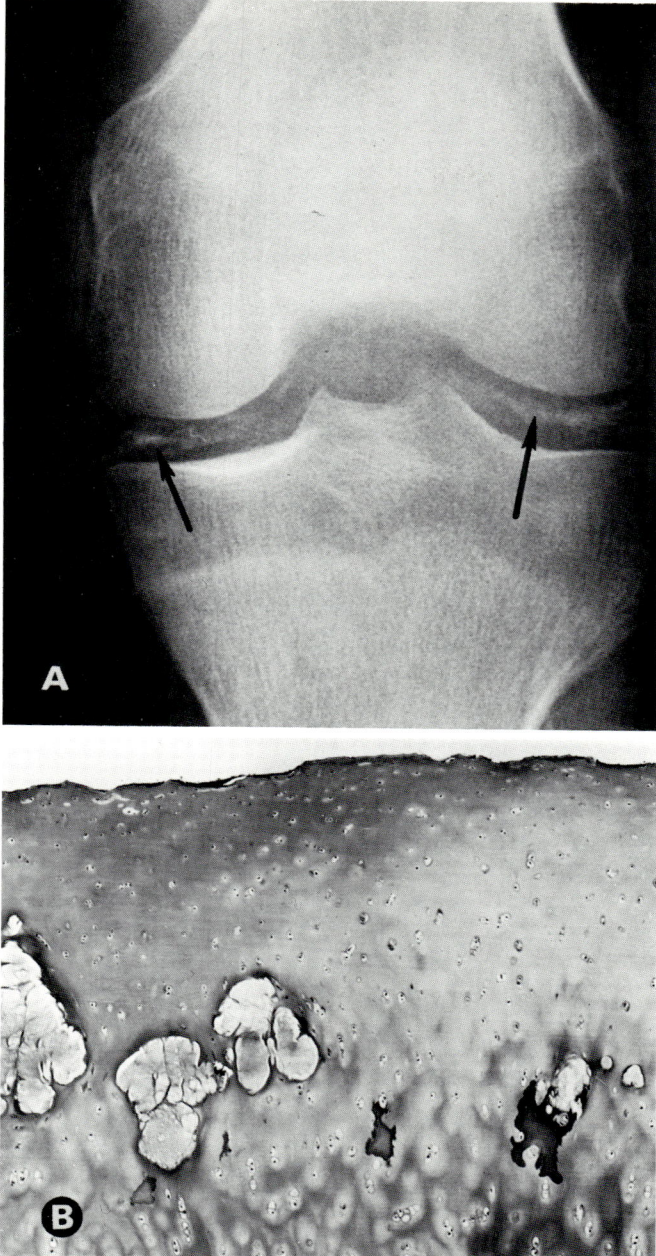

Figure 12.9. Articular chondrocalcinosis. *A.* The roentgenographic shadow of the calcium pyrophosphate (*arrows*) is located in the menisci. *B.* The crystals were present originally in the cystic areas of degeneration of the cartilage but were removed during preparation of the section for technical reasons. (Hematoxylin and eosin, ×33.)

femur. The circulatory embarrassment is identified in only a minority of cases, *e.g.*, in fractures of the femoral neck which interrupt the retinacular vessels. Another example is caisson disease, an occupational disorder of tunnel workers who may be subjected to rapid decompression. When blood gases come out of solution too rapidly, they form air emboli. After many years, multiple infarcts of the ends of the bones appear even if the victims have never experienced the acute painful episodes of decompression (the bends). In most cases of aseptic necrosis, no vascular occlusion is demonstrated and compression of intramedullary blood vessels against the rigid bony casing by swollen or hyperplastic marrow tissues seems the more likely mechanism. Aseptic necrosis is thus a fairly frequent complication of hematopoietic disorders such as leukemias, sickle cell anemia, and Gaucher's disease. The source of the increased medullary pressure in the aseptic necrosis that so often complicates prolonged corticosteroid therapy of systemic lupus erythematous or rheumatoid arthritis is less clear. Perhaps it is a Cushingoid enlargement

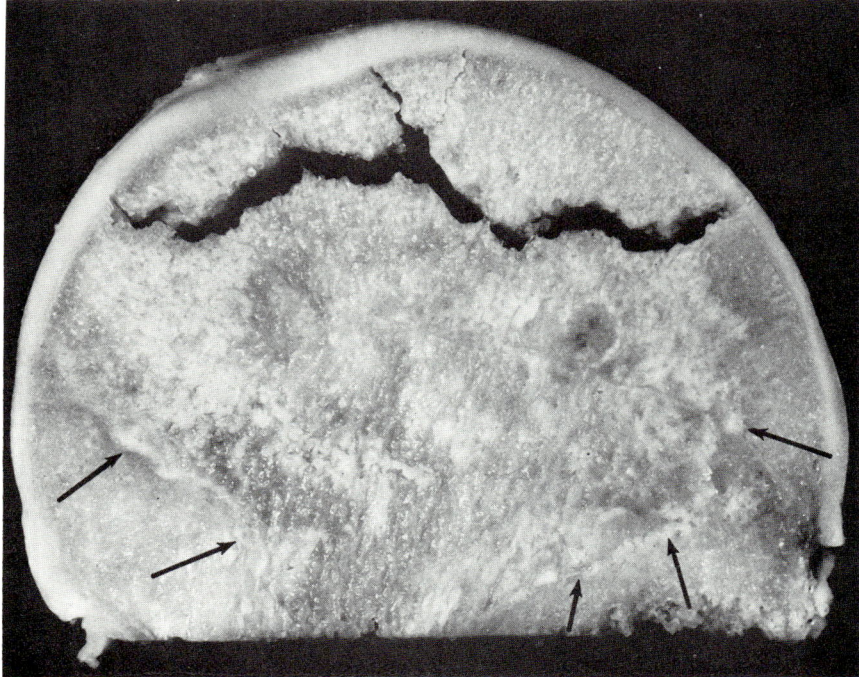

Figure 12.10. Aseptic necrosis, head of femur. The *arrows* point to the margin of the wedge-shaped infarct. A subchondral fracture is present but there is no collapse yet and the cartilage is intact.

of the adipose tissue cells in the marrow. Similar difficulties exist in understanding those cases associated with alcoholism and gout, and idiopathic instances.

Infarction of bone in itself produces relatively few symptoms. There is little resorption of the necrotic bone except at its marginal interface with

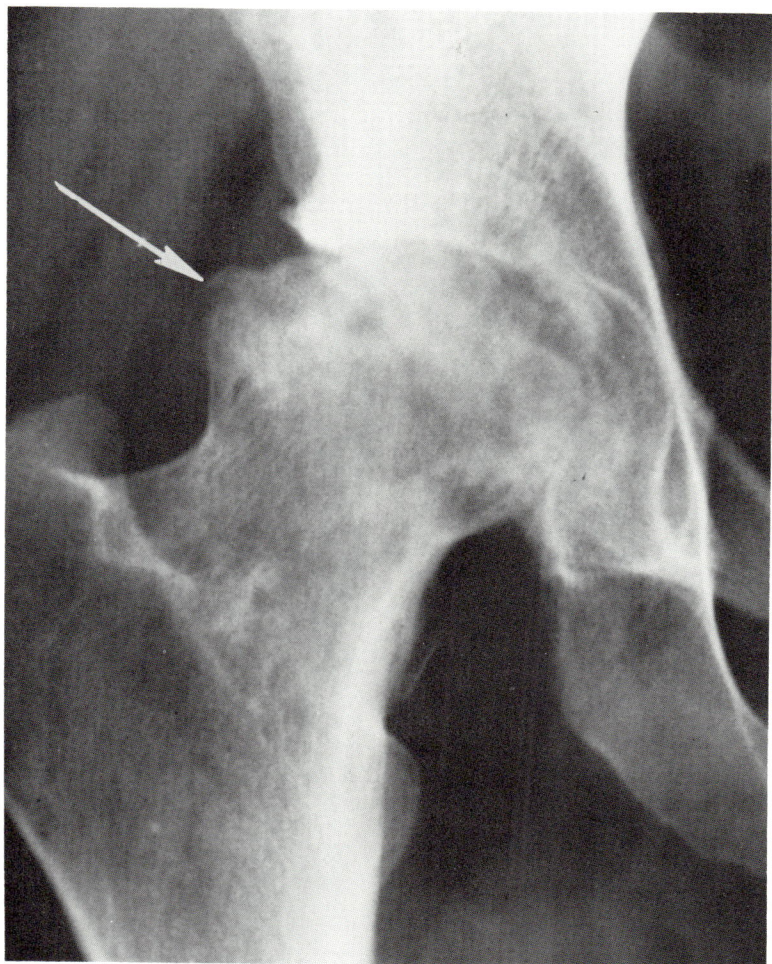

Figure 12.11. Aseptic necrosis of 2 years' duration, right femoral head of a 59-year-old alcoholic. The femoral head has lost its normal globoid contour and trabecular spray pattern. The superior surface is irregular and somewhat flattened. Patchy areas of radiolucence and sclerosis extend from the head into the neck. Secondary osteoarthritis has supervened as indicated by the loss of the joint "space" and the presence of osteophytes (*arrow*).

viable tissue. A slow replacement process called creeping substitution takes place if the lesion is small. The articular cartilage overlying infarcted bone remains viable for long periods of time because it draws its nourishment from the synovial fluid rather than the bone. The principal complaints arise when the weakened dead bone becomes fractured (Fig. 12.10). Osteoporosis induced by corticosteroid compounds facilitates the fracture. Progressive collapse of the joint surface results in pain and difficulty of motion. X-ray films disclose the collapse, reactive sclerosis of the adjacent bone, focal osteoporosis, and superimposed osteoarthritic deformity (Fig. 12.11).

The juvenile osteochondroses are special forms of idiopathic aseptic necrosis that occur with some frequency in the epiphyses of young children. They usually are called by a series of confusing eponyms. The most important of the osteochondroses is Legg-Perthes disease which affects the epiphysis of the head of the femur. It occurs during the first decade of life and causes a limp. Secondary osteoarthritis may be a late sequel.

REFERENCES

Amako, T., Kawashima, M., Torisu, T., and Hayashi, K. Bone and joint lesions in decompression sickness. Semin. Arth. Rheum. 4:151, 1974.

Bruckner, F. E. and Howell, A. Neuropathic joints. Semin. Arth. Rheum. 2:47, 1972.

Byers, P. D., Contepomi, C. A., and Farkas, T. A. A *post-mortem* study of the hip joint including the prevalence of the features of the right side. Ann. Rheum. Dis. 29:15, 1970.

Duthie. R. B., Matthews, J. M., Rizza, C. R., and Steel, W. M. The Management of Musculo-Skeletal Problems in the Haemophilias. Oxford, Blackwell Scientific, 1972.

Herndon, J. H., and Aufranc, O. E. Avascular necrosis of the femoral head in the adult. Clin. Orthop. 86:43, 1972.

Marmor, L., and Peter, J. B. Osteoarthritis of the hand. Clin. Orthop. 64:164, 1969.

McDevitt, C. A. Biochemistry of articular cartilage. Nature of proteoglycans and collagen of articular cartilage and their role in ageing and osteoarthrosis. Ann. Rheum. Dis. 32:364, 1973.

Seegmiller, J. E. Metabolic aberrations in gout. Clin. Orthop. 71:87, 1970.

van der Korst, J. K., Geerards, J., and Driessens, F. C. M. A hereditary type of idiopathic articular chondrocalcinosis. Survey of a pedigree. Am. J. Med. 56:307, 1974.

Woolf, C. M., Koehn, J. H., and Coleman, S. S. Congenital hip disease in Utah: the influence of genetic and nongenetic factors. Am. J. Hum. Genet. 20:430, 1968.

13

BACKACHE

About 80 per cent of us suffer from backache at some time in our lives. Several causes have already been indicated—osteoporosis, tumors, and vertebral osteomyelitis among them. There are additional common spinal diseases in which the etiology is known, but for many the precise pathologic mechanisms are not known. For this reason, loose terms such as lumbosacral strain (lumbago) and sciatica retain a distinct clinical usefulness.

STRUCTURE AND FUNCTION OF THE SPINE

To maintain posture, muscles exert large forces on the vertebral column. This is particularly true in its lower portions where, through their leverage, the pressure on the anulus fibrosus is of the order of 1000 pounds per square inch. These pressures are greatly increased by bending, lifting, and even sneezing, actions that often trigger acute back problems. The pressure is 30 percent lower in the standing than the sitting posture and still lower when reclining.

Each vertebra, rib articulations excluded, has two systems of joints: the intervertebral discs, which are amphiarthroses, and the apophyseal (facet, posterior) joints which are diarthroses (Fig. 13.1). The general physiology of the discs has been considered in Chapter 3. They support the bulk of the force acting on the thoracolumbar portions of the column, while the diarthroses are more important in the cervical region where mobility is greater. A unique diarthrosis, the atlantoaxial joint, permits rotary motion of the head about the odontoid process that juts superiorly from the axis. The sacroiliac joints also are diarthroses but have hardly any motion because they are constrained by strong fibrous ligaments. The vertebral bodies are supported by collagenous anterior and posterior longitudinal ligaments. In the ligamenta flava that connect the laminae of the neural arches, and the dorsal supraspinous ligaments, the dominant fibrous protein is elastin.

Back pain has two distinct qualities depending on its source of origin. *Somatic* pain arises from irritation of nerves in the periosteum and

146   *Musculoskeletal System*

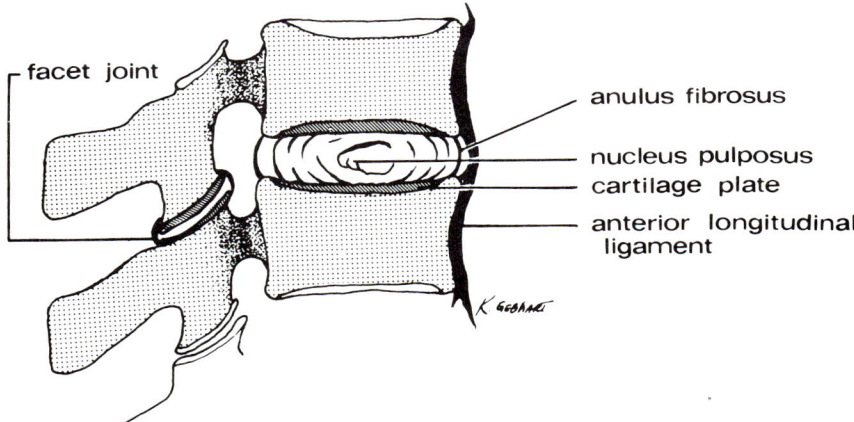

Figure 13.1. Vertebrae articulate with each other through two separate structures: the intervertebral disc and the apophyseal (facet) joints.

ligaments. It has a deep-seated, ill-defined, and aching character. When severe, somatic pain causes a feeling of sickness and is accompanied by nausea, vomiting, and sweating. The pain radiates in a sclerotome fashion and often is referred to a remote site. A lesion of a thoracic vertebra may, for example, come to simulate a myocardial infarction. *Neuralgic* pain results from compression or other stimulation of nerve roots by lesions of the bony wall of the neural foramina or adjacent supporting tissues. It follows the superficial or dermatome representation. The pain is sharp and lancinating and often is accompanied by paresthesias. When root injury is severe and of long duration, major motor as well as sensory losses result.

Reflex spasm of paravertebral muscles is conspicuous in many cases of backache and is not related to nerve root irritation. It comes on acutely and may result in bizarre twisting of the spine. The spastic muscles are tender and palpably hardened. To relieve the spasm while avoiding generalized loss of muscle contraction, a class of muscle relaxants different from the curariform and acetylcholine analogues is useful. These act on the central nervous system rather than the neuromuscular junction. The most widely employed compounds are methocarbamol (Robaxin) and carisoprodal (Soma).

ANKYLOSING SPONDYLITIS (MARIE-STRÜMPELL DISEASE)

This is predominantly a disease of young men and accounts for about 20 percent of chronic backache in this group. The onset usually occurs before 30 years and 8 men are affected for every woman. Ankylosing spondylitis is a chronic progressive disease of both the amphiarthrodial and diarthrodial joints including the sacroiliac articulations. Joints of the limbs, particularly those close to the spine such as the hip, also are affected in one-fourth of patients. The disorder resembles rheumatoid arthritis in certain respects and it still is sometimes called rheumatoid spondylitis. Although cervical diarthrodial joints often are involved in seropositive rheumatoid arthritis (Fig. 13.2), there are many differences between that condition and Marie-Strümpell disease and it is now believed that these are two unrelated entities.

The pathologic process in the diarthroses resembles that of rheumatoid arthritis but there is a much greater tendency to go on to bony ankylosis. The anatomic events at the intervertebral discs are less well documented. The margins of the vertebrae to which the anulus fibrosus is attached first undergo osteolytic resorption and there follows a bony transformation of the anulus. As a result, the vertebral bodies become fused into an immobile "poker back." It is the ossification of the anulus fibrosus that is responsible for the "bamboo spine" appearance on

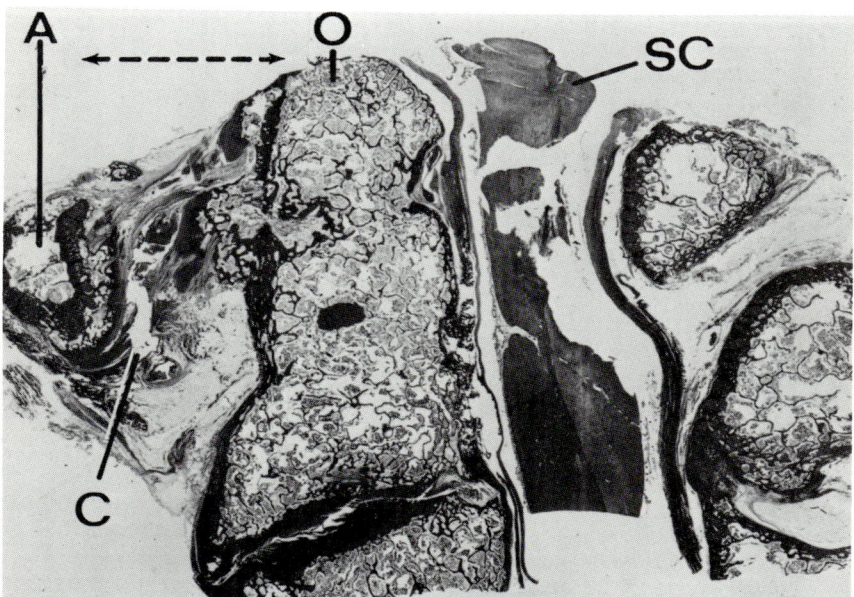

Figure 13.2. Sagittal section of severely subluxated atlantoaxial joint in rheumatoid arthritis. The *dotted line* indicates the extent of the separation of the original articular surfaces of the anterior arch of the atlas (*A*) and the odontoid process (*O*). The joint surfaces are deformed and reduplicated. The subluxation has resulted from the postinflammatory gap in the capsule (*C*) and transverse ligament. Note the narrowing of the spinal cord (*SC*). Masson, ×1.5.)

148 *Musculoskeletal System*

x-ray films (Fig. 13.3). The intervertebral discs are not ordinarily narrowed although this may occur in advanced cases. Other spinal ligaments, collagenous and elastic, also may be involved, but the fact that the bamboo spine is seen in anteroposterior x-ray films shows that it is the anulus fibrosus which is the principal tissue affected: the spine has anterior and posterior but no lateral longitudinal ligaments. Sacroiliac ankylosis is a consistent feature and is manifested radiologically by bony obliteration of the joint space (Fig. 13.4). Ossification of the tendinous

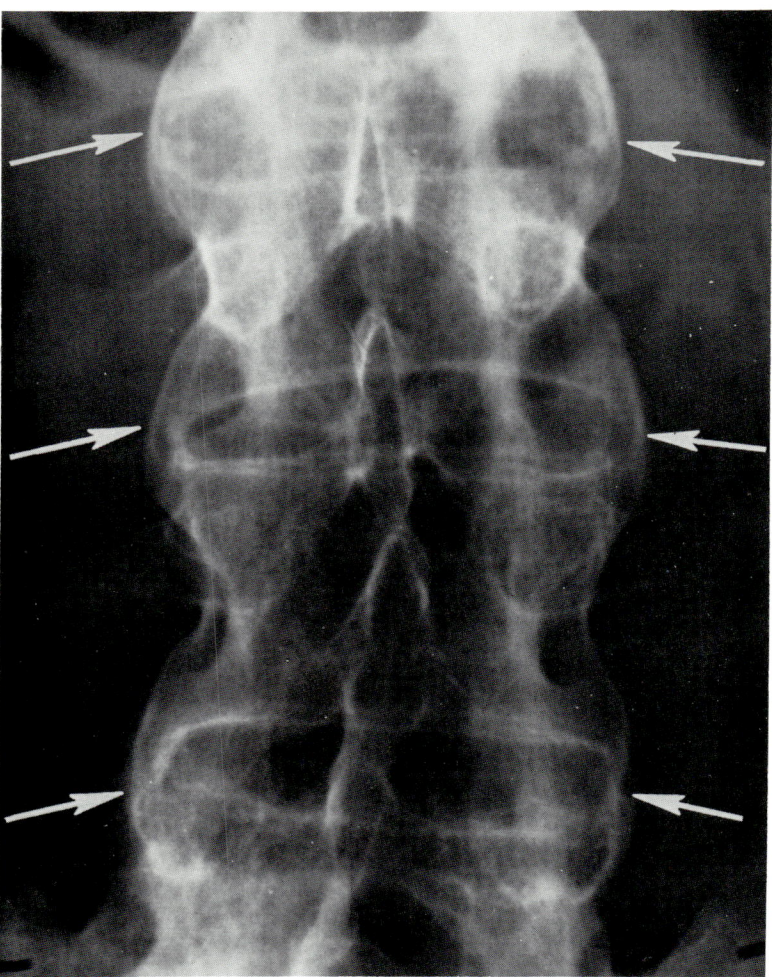

Figure 13.3. Bamboo spine in ankylosing spondylitis. The *arrows* point to the bony bridges uniting the vertebral bodies at the periphery of the intervertebral discs.

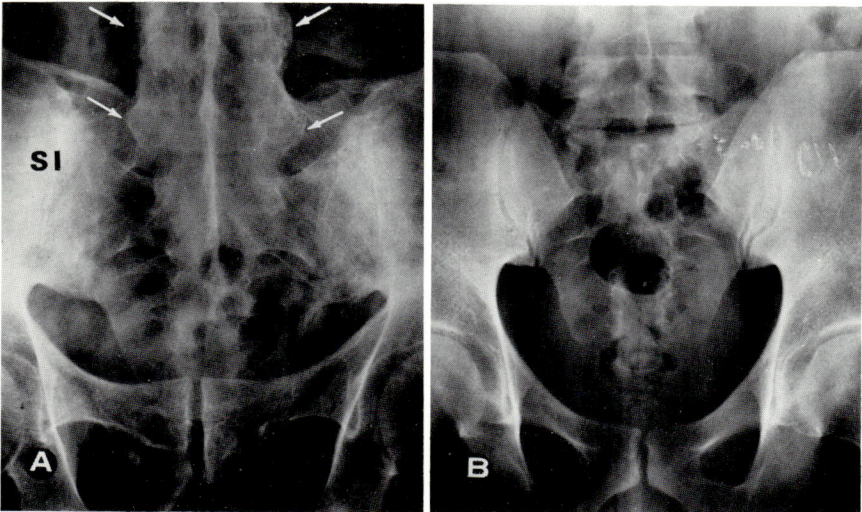

Figure 13.4 A. Sacroiliac ankylosis in Marie-Strümpell disease contrasted with normal (*B*). The bony ankylosis is manifested by the absence of a radiolucent "joint space" in the position of the sacroiliac articulation (*SI*). Note the bamboo spine (*arrows*).

attachments to bones (enthesopathies), *e.g.*, along the inferior margins of the pelvis, is another useful x-ray finding.

The backache usually begins insidiously in the sacroiliac area. It has a variable character but ultimately becomes constant, dull, and aching. After many years, pain yields to loss of motion of the spine. Unless proper exercises and other preventive measures are employed, the vertebral column tends to become fixed in a severe kyphotic position and the head cannot be moved. Attempts to splint such patients intensifies the development of the rigidity. Immobilization thus is an important factor in the process of ossification.

Ankylosing spondylitis is a systemic disease. Its etiology is unknown except for the fact that more than 90 percent of affected persons are of the HL-A W27 histocompatibility type. The disease therefore requires a certain genetic substrate. Serologic tests for rheumatoid arthritis are negative and there are no subcutaneous nodules. Iritis occurs in about 25 percent of cases and a smaller proportion develop insufficiency of the aortic valve. The valvular lesions grossly resemble those of syphilis insofar as they come about through a weakening of the aortic ring and separation of the commissures.

In past years, x-irradiation of the spine was employed in the treatment of Marie-Strümpell disease. An unusually high prevalence of leukemia in

DEGENERATIVE DISEASES OF THE SPINE

Degenerative disease takes two forms in the spine depending on which of its articular systems is involved. Spondylosis is the type that occurs in the intervertebral disc while spinal osteoarthritis affects the diarthrodial joints. The two processes commonly coexist, although within a given vertebra only one or the other structure is affected.

Degenerative changes in the nucleus pulposus result in a loss of its hydrostatic pressure-distributing properties. As a result, the discs collapse, one of the major reasons people become shorter when they grow old. This is a prelude to fissuring of the hyaline cartilage plate that lies between the disc and the cortex of the vertebral bodies, and it leads in time to the sequences of eburnation and marginal osteophytosis. Osteophytes ordinarily are situated on the anterolateral surfaces of the vertebral body, particularly in regions of maximum curvature where pressures are highest. In the cervical spine, osteophytes often protrude posteriorly toward the spinal canal. They thus compromise the circulation of the spinal cord and result in major neurologic defects. Posterior cervical osteophytes are important factors in causing death following automobile whiplash injuries in the elderly. Osteophytes also often impinge on the neural foramina of the cervical spine close to the uncinate processes and are a frequent cause of brachial neuralgias., Because lumbar osteophytes are so common in the older population, compensation courts grapple every day with the problem of determining whether they have anything to do with any particular case of backache.

Displacement or "herniation" of the nucleus pulposus is a common and disabling condition that occurs in young as well as middle-aged adults. It usually affects only one or two discs, mostly in the lumbosacral region but sometimes in the neck instead. The nucleus pulposus is partially extruded through a tear in the posterior part of the anulus fibrosus and impinges on the nerve roots. The displaced nucleus pulposus can be located radiologically by myelography, a procedure in which a radiopaque contrast medium is injected into the spinal fluid. Sciatica develops as a consequence of the root irritation, and is manifested by radiation of low back pain into the lower extremity. The herniation typically is precipitated by a specific exertion and the patient may be aware of a snapping sensation. The exertion is often disproportionately small and one presumes that there must be some underlying abnormality in the anulus. The late structural and chemical changes in the discs are identical to those of spondylosis albeit confined to the affected segment. The herniation often is reduced spontaneously or following the use of spinal traction and bedrest to diminish the compressive stress on the disc. Protracted failure to correct itself may require surgical excision of the degenerated disc tissue.

Schmorl's nodes are somewhat analogous to the preceding but the displacement of the disc substance takes place into the cancellous bone of the vertebral body rather than posteriorly. These are common findings in x-ray films and usually are asymptomatic.

Developmental and structural anomalies of many sorts cause "bad backs." Scheuermann's disease is a juvenile osteochondrosis of thoracic vertebrae. There are several types of spondylolisthesis, an anterior displacement of one vertebral body upon the next below. Idiopathic scoliosis, a relatively frequent disorder of adolescent children, may lead to severe lateral curvatures of the spine but in only a minority of cases is pain a major complaint.

When all is said and done, there remain many backaches for which none of the above explanations is applicable. In many cases, some ill-defined, mechanically derived tears ("strains") in the supporting connective tissues are presumed to be at fault. In such cases, mechanical measures—a mixture of rest and muscle strengthening exercises, avoidance of stress, use of braces and corsets—and analgesics afford a measure of relief. Sometimes the source of the backache is to be sought not in the spinal column but in a visceral lesion, *e.g.*, an aneurysm of the abdominal aorta or in a weakness of the pelvic floor in parous women. Finally, it is important to distinguish that part of the backache population which is of psychosomatic origin or malingering.

REFERENCES

BALL, J. Enthesopathy of rheumatoid and ankylosing spondylitis. Ann. Rheum. Dis. 30:213, 1971.

BLAND, J. H. Rheumatoid arthritis of the cervical spine. J. Rheum. 1:319, 1974.

BROWN, W. M. C., AND DOLL, R. Mortality from cancer and other causes after radiotherapy for ankylosing spondylitis. Br. Med. J. 2:1327, 1965.

COLLIS, D. K., AND PONSETI, I. V. Long-term follow-up of patients with idiopathic scoliosis not treated surgically. J. Bone Joint Surg. 51A:425, 1969.

FINNESON, B. E. Low Back Pain. Philadelphia, J. B. Lippincott, 1973.

JACKSON, R. The Cervical Syndrome, 3rd ed, revised, Springfield, Ill., Charles C Thomas, 1971.

WILKINSON, M. (Ed.) Cervical Spondylosis. Its Early Diagnosis and Treatment, 2nd ed. Philadelphia, W. B. Saunders, 1971.

14

LABORATORY DATA IN MUSCULOSKELETAL DISEASES

Laboratory tests related to the pathologic events considered in the preceding chapters play an important part in diagnosing and assessing the course of many musculoskeletal diseases. We present here four tables that summarize commonly employed biochemical and immunologic procedures; and a fifth on the classification of synovial fluids, according to their properties, that currently enjoys much clinical favor. It is apparent that hardly any of the findings is pathognomonic. Additional specific diagnostic tests, *e.g.*, culture and crystal identification, have been indicated previously in the text. The student should be familiar with these tables for the limitations as well as the usefulness of the tests, and so place proper weight on them in evaluating both their clinical and their pathogenetic significance in this large branch of medicine. It also gives

TABLE 14.1
Laboratory Values in Musculoskeletal Diseases: Serum Enzymes

Enzyme*	Normal Value (International Units)	Clinical Disorders Associated with Elevated Values
Alkaline phosphatase	50–150 (childhood) 25–85 (adult)	Paget's disease Rickets and osteomalacia Hyperparathyroidism Metastatic bone tumors (Liver disease, heat-stable)
Acid phosphatase	1–2.6	Metastatic carcinoma of prostate Gaucher's disease
CPK	7–45	Active muscular dystrophy, polymyositis
Aldolase	0.5–7.9	(Recent myocardial infarction)
SGOT	9–22	Active muscular dystrophy, polymyositis
SGPT	3–23	(Recent myocardial infarction)
LDH	62–134	(Hepatocellular disease)

*CPK, creatine phosphokinase; SGOT, serum glutamic-oxaloacetic transaminase; SGPT, serum glutamic-pyruvic transaminase; LDH, lactate dehydrogenase.

TABLE 14.2.
Laboratory Values in Musculoskeletal Diseases: Serum Electrolytes and Small Molecules

Substance	Normal Values	Clinical Disorders Associated with	
		Elevated values	Reduced values
Potassium	3.5–5.5 meq/liter	Severe myoglobinurias	Familial periodic paralysis
Calcium	8.5–10.5 mg/100 ml	Primary hyperparathyroidism, Metastatic bone tumors Multiple myeloma (Sarcoidosis)	Hypoparathyroidism (Chronic renal failure)
Phosphorus	3.2–4.6 mg/100 ml	(Chronic renal failure)	Primary hyperparathyroidism Hypophosphatemic rickets and osteomalacias
Urate (low purine diet)	5–7.5 mg/100 ml (colorimetric) 4.5–6.5 mg/100 ml (enzymatic) 7.5–8.4 mg/100 ml (automated)	Gout (Genetic) (Myeloproliferative disorders) (Metabolic acidosis: starvation, diabetic, alcoholic, glycogen storage diseases) (Aspirin, low dosage) (Chronic renal failure) (Thiazide diuretics) (Hypertension)	(Aspirin, high dosage)

TABLE 14.3.
Laboratory Values in Musculoskeletal Diseases: Urine

Substance	Normal Value	Clinical Disorders Associated with	
		Elevated value	Reduced value
Calcium (low Ca diet)	150–200 mg/24 hr	Primary hyperparathyroidism Metastatic bone tumors	Renal failure
Phosphorus clearance (Clp)*	6.3–15.5 ml/min	Hyperparathyroidism Hypophosphatemic rickets or osteomalacia	
Urate (low purine diet)	300–600 mg/24 hr	Over-producer gout Myeloproliferative disorders	
Mucopolysaccharide (as uronic acid)		Morquio's disease Hurler's syndrome	
Children	12–36 mg/24 hr		
Adults	3–9 mg/24 hr		
Hydroxyproline (low collagen diet)	15–35 mg/24 hr	Hyperparathyroidism Paget's disease	
Creatine:creatinine ratio	6%	End-stage muscular dystrophy and polymyositis (Starvation)	

* $\text{Clp} = \dfrac{\text{Conc. urinary phosphorus}}{\text{Conc. serum phosphorus}} \times \text{ml urine/min.}$

without saying that normal values for all of these tests must be established in each laboratory performing them.

REFERENCE

COHEN, A. S. (Ed.) Laboratory Diagnostic Procedures in the Rheumatic Diseases. Boston, Little Brown, 1967.

TABLE 14.4.
Serologic Abnormalities in Rheumatic Diseases

Serologic Test	Normal or Upper Equivocal Value	Frequency of Positive Tests in Various Diseases (%)							
		Rheumatoid arthritis	Rheumatic fever	Systemic lupus erythematosus	Marie-Strümpell	Scleroderma	Dermato- and polymyositis	Sjögren's syndrome	Miscellaneous
Rheumatoid factor	Latex fixation 1:160 Bentonite flocculation 1:32 Sensitized sheep RBC 1:32	70		40	5	30	25	90	Old age, sarcoidosis, chronic hepatitis, bacterial endocarditis, 20–30
LE Cell		20	5	90	5	5	8	20	Hydralazine, drug sensitivity, 25
Antinuclear antibody		25	0	98		40	30	70	Chronic active hepatitis, 40
Antistreptolysin O	150–250 units		75						
Hemolytic complement, serum	80–140 units	↑10	↑90	↓67		↓7			
Synovial fluid		↓81*							
False positive tests for syphilis				20					

* Seropositive cases only; ↑ = increased; ↓ = decreased.

TABLE 14.5.
Classification of Synovial Fluids

Property	Normal	Group I	Group II	Group III	Group IV
Appearance	Clear, straw color	Clear or opalescent straw color	Turbid, yellow	Turbid, yellow-orange	Bloody or xanthochromic
Viscosity	High	High	Low	Low	Variable
Fibrin clot	Absent	Usually absent	Present	Present	Usually absent
Protein (g/100 ml)	1.7–2.1	2.5–3.3	2.5–6.0	3.5–7	3.0–4.0
Mucin clot test*	Good	Good	Fair to poor	Poor	Variable
Nucleated cells/mm³	200	200–500	2,000–100,000	20,000–200,000	200–10,000
% Polymorphonuclear leukocytes	25	25	50	75	50
Plasma-synovial fluid glucose concentration difference (mg/100 ml)	10	10	25	50	25
Diseases usually associated	None	Non-inflammatory (Osteoarthritis, neuropathic, traumatic)	Inflammatory, Non-Infectious (Rheumatoid, gout, pseudogout, psoriatic)	Infectious (Septic)	Hemorrhagic (Hemophilic, traumatic with hemorrhage, pigmented villonodular synovitis)

* See Figure 3.3.

INDEX

Acetylcholine, 50, 70, 146
Achondroplasia, 75, 82
Acid phosphatase, 18, 95, 96, 152
Acidosis, 14, 94, 137, 140, 153
Acromegaly, 10, 79
Actomyosin, 44, 45
Adenosine triphosphatase, 46, 47
Adenosine triphosphate, 27, 46-48
Agammaglobulinemia, 122, 123
Aging, 7, 129
Aldolase, 48, 62, 69, 152
Alkaline phosphatase, 17, 27, 140
   serum, 27, 33, 82, 85, 92, 94, 95, 98, 101, 152
Amphiarthrosis, 35, 41-42, 145, 147
Androgen, 31, 37, 78, 95
Anemia (*see also* Sickle cell anemia), 89, 101, 115, 124
Ankylosing spondylitis, 121, 122, 146-150, 155
Ankylosis, 115, 116, 120, 128, 133, 147, 148
Anti-DNA antibodies, 71, 124
Antinuclear antibodies, 124, 126, 127, 155
Antistreptolysin O, 127, 155
Anulus fibrosus, 41, 42, 146-148, 150
Arteritis (*see also* Polyarteritis nodosa), 64, 121, 126
Arthralgia, 112, 114, 125, 126
Arthritis (*see also* Rheumatoid arthritis), 83, 84, 114, 129, 138
   infectious, 21, 109-113, 156
Arthrogryposis multiplex congenita, 39
Aseptic necrosis (*see also* Bone infarcts), 96, 140, 142-144

Backache, 85, 92, 94, 144-151
Bamboo-spine, 147-149
Biopsy, bone, 11, 24, 26, 85, 97, 107
   joint, 111
   muscle, 46, 50, 56, 57, 60, 62, 66
Birefringence, collagen, 10, 12, 13, 38
   crystals, 139, 140
   polarized light, 10, 11, 37
Bone age, 21, 75, 76, 77, 79
Bone infarcts (*see also* Aseptic necrosis), 95, 140, 142-144
Bone marrow (*see also* Hematopoiesis), 20, 37, 89, 95, 102, 103, 105, 111, 130
Bouchard node, 131

Brodie's abscess, 108
Brucella, 108
Bursa, 18, 112, 114
Bursitis, 112, 114

Caffey's disease, 103, 104
Caisson disease, 142
Calcitonin, 28-30, 86, 100
Calcium, 5, 12, 14-16, 24, 28-30, 47, 81, 83, 86, 88
   balance, 22, 24, 32
   ionized, 13, 15, 27, 28, 33, 43, 45, 46, 89, 94
   pyrophosphate, 27, 139-141
   serum, 27, 28, 33, 85, 92, 94, 95, 98, 101, 153
   urine, 32, 93, 94, 154
Callus, 16, 31, 32, 82, 88
Canaliculi, 11, 14, 17, 22
Cartilage (*see also* Fibrocartilage), 1, 2, 7, 9, 10, 16, 32, 39, 63, 73, 109, 111, 132, 138
   articular, 4, 5, 35-41, 79, 94, 111, 115, 117, 128-130, 133, 138, 140, 141, 144
   calcified, 11, 16, 21, 22, 27, 35, 38, 89, 100, 102
   vertebral, 41, 146, 150
Cement line, 3, 14, 97, 100
Charcot joint (*see* Neuropathic arthropathy)
Cholinesterase, 50, 70, 71
Chondrocalcinosis (*see also* Pseudogout), 27, 140, 141
Chondrocytes, 16, 17, 21, 31, 37, 77, 89, 109, 129
Chondroitin sulfate, 8, 9, 11, 37, 77
Circulation, blood, 121, 125, 150
   bone, 23, 25, 26, 95, 98, 102, 105, 106, 110, 140, 142
   intervertebral disc, 41
   joint, 37, 39, 129, 130
   Paget's disease, 97, 98
Clubbing of fingers, 101
Coccidioidomycosis, 112
Codfish vertebra, 85, 87, 89, 91
Colchicine, 138, 140
Collagen, 1-11, 13, 16, 18, 20, 27, 33, 41, 66, 81
   cartilage, 4-6, 37, 38, 111, 128, 129
   ligament, 145, 148

Collagen diseases, 114, 123–127
Collagenase, 4, 6, 60, 108, 115
Complement, 39, 111, 113, 118, 119, 124, 155
Contracture, 59, 61, 68, 69
Corticosteroid (see also Hyperadrenalcorticism), 31, 64, 67, 71, 72, 78, 95, 103, 106, 108, 119
  bone infarct, 142
  osteoporosis (see Osteoporosis, corticosteroid)
CPK, 48, 49, 62, 69, 152
Cramps, 58, 59
Creatine, 48, 49, 69, 70, 154
Cysticercus cellulosae, 60

Deafness, 98, 109
Degenerative joint disease (see also Osteoarthritis), 77, 114
Dermatan sulfate, 9, 77
Dermatomyositis, 61–63, 124, 155
Desert rheumatism, 112
Diabetes mellitus, 78, 108, 132, 139, 153
Diarthrosis, 35–41, 111, 145, 147
Distal muscle syndromes, 67
Dwarfism, 10, 73, 75, 78, 89
Dysplasia, hip, 130
  metaphyseal, 21
  multiple epiphyseal, 75

Eburnation, 128–131, 150
Echinococcus, 109
Elasticity, viscoelasticity, 2, 3, 47, 54, 100
  Young's modulus, 2, 11, 88
Elastin, 1, 10, 79, 145, 148
Electromyography, 43, 52–56, 58, 59, 62, 64, 66, 70
Endochondral ossification, 6, 20, 21, 75, 100, 109
Endomysium, 44, 51, 56, 59, 60, 63, 65, 66, 69, 71
Endosteum, 22, 23, 25, 85, 92
Eosinophilia, 60, 109
Epiphysis, 15, 21, 30, 39, 73, 75, 89, 109, 140, 144
  secondary centers, 21, 79, 90, 91
  union, 25, 26, 31, 73, 76–79
Estrogens, 31, 86

Fanconi syndrome, 89, 91
Fatigue, 32, 49, 66, 70, 115
Fibrinoid, 120, 121, 123, 124, 126
Fibrocartilage, 9, 10, 16, 35, 39
Fluoride, 15, 104
Fracture, 2, 3, 11, 31, 32, 81–84, 88, 89, 94, 96, 97, 100, 101, 105, 109, 144
  femoral neck, 85, 92, 94, 142
  microfracture, 128, 130, 132

Gas gangrene, 6, 59, 60
Gaucher's disease, 94, 96, 142, 152
Geographic variation, 60, 94, 97, 109, 114, 130
Gigantism, 79
Glycogen, 46–48
  storage diseases, 72, 153
Glycolysis, 16, 37, 46, 48
Gonococcus, 111, 112
Gout, 64, 114, 133–140, 143, 153, 154, 156
Ground substance, 1, 2, 5, 7–11, 13, 16, 27, 37, 41, 111
Growth arrest lines, 74
Growth disturbances, 21, 73–80, 96, 109
Growth hormone, 30, 31, 77–79
Guillain-Barré syndrome, 68

Haversian canal, 12, 23–26
Heberden's node, 130, 131
Hematopoiesis, 89, 95, 102, 106, 142
Hemophilic arthropathy, 132, 156
Hepatitis, 112, 113, 124, 152, 155
Hip joint, 75, 107, 129, 130, 140, 142–144, 147
Histocompatibility antigen, 121, 149
Homocystinuria, 6
Hurler's disease, 77, 154
Hyaluronic acid, 8, 9, 11, 37, 39, 40
Hydralazine sensitivity, 124, 155
Hydroxyapatite, 11, 12, 27, 140
Hydroxyproline, 4, 6, 28, 29, 33, 94, 154
Hyperadrenalcorticism (see also Corticosteroids), 31, 63
Hypercalcemia, 30, 32, 33, 92, 94, 95
Hyperkalemia (see also Potassium), 64
Hyperostosis, 91, 97–104
Hyperparathyroidism, 17, 33, 86, 92–94, 152–154
Hyperthyroidism, 72
Hyperuricemia, 136, 138, 140
Hypocalcemia, 30, 33, 58, 89, 92
Hypophosphatemia, 89, 91–93, 153, 154
Hypothyroidism, 76, 78

Immunoglobulin, 117, 119, 122, 123
Inheritance, bone disease, 75, 77, 79, 81, 82
  chondrocalcinosis, 140
  Gaucher's disease, 95, 96
  gout, 138
  hemophilia, 132
  Lesch-Nyhan syndrome, 136
  mucopolysaccharidoses, 8
  muscle disease, 59, 68–70
  ochronosis, 131
  rheumatoid arthritis, 121

Innervation, bone, 25, 26, 145
  joint, 41, 51
  muscle, 41, 43, 46, 52, 54, 55
Intervertebral disc, 35, 41, 42, 85, 87, 106, 108, 145
  ankylosing spondylitis, 147, 148
  degeneration, 91, 132, 150-151
Intestine, 28-30, 32, 60, 137
  diseases, 63, 89, 122, 125
Involucrum, 106, 107
Iritis, 121, 144

Juvenile rheumatoid arthritis (*see also* Still's disease), 73, 112, 121, 122

Keratan sulfate, 8, 9, 37, 77
Kidney (*see also* Renal, Uremia), 28-30, 89, 131
Kyphos, 73, 79, 85, 108, 149

Lactate, 37, 48, 58, 59
LDH, 48, 152
Lamellar bone, 11-13, 21, 81, 88
LE cell, 124, 125, 155
Lead poisoning, 15, 68
Legg-Perthes disease, 144
Lesch-Nyhan syndrome, 136
Ligaments, 1, 4, 39, 41, 104, 112, 116, 132, 135
  spinal, 145-148
Liver (*see also* Hepatitis), 28-30, 49, 117, 131, 152
Looser lines, 88, 89
Lumbago, 145
Lymphorrhage, 51, 59, 71
Lysosomes, 6, 8, 16, 18, 37, 96, 115, 118, 119, 138
Lysozyme, 8

Malabsorption syndrome, 89, 91, 92
Marfan's syndrome, 6, 79
Marie-Strümpell disease (*see* Ankylosing spondylitis)
Matrix vesicle, 16, 27
McArdle's disease, 59, 72
Membranous bone, 13, 20, 22, 63, 79
Meningococcus, 110
Menopause, 86, 130
Metachromasy, 11
Metastatic bone tumors, 94, 95, 97, 114, 152-154
Metastatic calcification, 33, 94
Metaphysis, 21, 26, 73, 90, 91, 108, 109
Mineralization, 5, 11, 16, 26-30, 87, 89, 104
Mitochondria, 16, 46, 48
Morquio-Brailsford disease, 75, 154

Mucopolysaccharides, 7-9, 11, 37, 154
Mucopolysaccharidosis, 8, 75, 76, 95
Muscle, 43-57
  atrophy, 66, 67
  degeneration, 57, 61, 66, 68
  regeneration (*see* Regeneration, muscle)
  relaxants, 50, 146
  spasm, 41, 58, 146
  spindle, 50, 51
  typing, 46, 47, 52, 54, 67, 70
  weakness, 58, 61, 64, 66-73
Muscular dystrophies, 52, 61, 67-70, 114, 152, 154
Myalgia, 58
Myasthenia gravis, 51, 59, 70, 71
Myelosclerosis, 102, 103
Mycardial infarction, 49, 84, 146, 152
Myofiber, 43-47, 52, 54, 57-60, 66, 67, 70, 71
Myoglobin, 46, 48, 64, 65, 153
Myopathy, 49, 55, 61, 66, 70-72
Myositis, 59-63
  ossificans, 63, 64
Myotonia, 55, 56, 59, 67, 70

Nephrolithiasis, 92, 94
  uric acid, 139
Neuralgia, 146, 150
Neuromuscular junction, 43, 50, 52, 59, 70, 146
Neuron, 43, 51, 52, 58, 67, 96
Neuropathic arthropathy, 41, 132, 133, 156
Neuropathy, 49, 52, 54, 55, 66-68, 121, 126, 132
Newman's sheath, 11, 14
Nucleus pulposus, 9, 41, 42, 85, 91
  herniation, 42, 150, 151

Ochronosis, 131, 132
Osteitis, 105
Osteitis fibrosa, 17, 81, 82, 89, 92-94
Osteoarthritis, 75, 128-133, 138, 143, 144, 150, 156
Osteoarthropathy, hypertrophic, 101, 102
Osteoblast, 4, 13, 17, 21, 27, 31, 81, 87, 94
Osteochondritis, 105, 108
Osteochondrosis, 144, 151
Osteoclasts, 17, 18, 21-24, 28, 29, 31, 85, 88, 94, 97, 100, 106, 130
Osteocytes, 11, 16-18, 23-25, 27, 81
Osteocytic osteolysis, 24
Osteogenesis imperfecta, 81-83
Osteogenic sarcoma, 14, 63, 100
Osteoid, 11, 17, 20, 21, 31, 82, 87, 88, 90, 92, 94
Osteolathyrism, 6
Osteomalacia, 12, 27, 81, 82, 86-92, 94, 114, 152-154

Osteomyelitis, 25, 105–110, 112, 145
Osteone, 14, 23
Osteopenia, 3, 81–96
Osteopetrosis, 3, 100, 101
Osteophyte, 128–132, 134, 143, 150
Osteoporosis, 30, 81, 82–88, 92, 94, 95, 114, 116, 145
  corticosteroid, 86, 95, 144
  disuse, 83, 132, 133
  senile, 85, 86, 91

Paget's disease, 97–100, 105, 152, 154
Pain, 101
  bone, 26, 81, 85, 86, 92, 94, 97
  joint, 41, 84, 112, 114, 115, 129, 130, 132, 138, 149
  muscle, 58–61, 64, 65
  quality, 145, 146, 149
Pannus, 115, 117
Parathyroid hormone (*see also* Hyperparathyroidism), 28–30, 93
Periosteum, 20–23, 25, 26, 30, 31, 36, 63, 106–108, 128
  innervation, 145
  new bone formation, 20, 79, 95, 97, 101–103, 106, 111, 112
  periosteitis, 103, 109
Phosphorus, 27, 30
  clearance, 93, 154
  phosphate, 27, 30, 47, 88, 89, 93, 94
  serum, 30, 85, 98, 101, 153
Phosphorylase, 46, 48, 59, 72
Piezoelectricity, 3, 16, 25
Pluripotentiality, connective tissue, 16, 31, 37, 63, 111, 128
Poliomyelitis, 32, 67, 73
Polyarteritis nodosa, 124, 126
Polymyalgia rheumatica, 64, 126
Polymyositis, 61–63, 67, 72, 114, 152, 155
Potassium (*see also* Hyperkalemia), 43, 58, 71, 153
Pressure (*see also* Stress), 13, 25, 35, 37, 41, 73, 130, 142, 145, 150
  swelling, 7, 37, 41, 85, 87, 91
Prostaglandin E_2, 30, 95
Proteoglycan, 2, 7, 20, 27, 37, 41, 129
Proximal muscle syndromes, 67, 70, 72
Pseudarthrosis, 31, 32
Pseudocyst, 128–131, 138
Pseudogout (*see also* Chondrocalcinosis), 139, 140, 156
Psoriatic arthritis, 106, 122, 156

Raynaud's phenomenon, 63, 125
Reflex dystrophy, 41, 84

Regeneration, bone, 31, 32
  cartilage, 37, 129
  muscle, 46, 51, 52, 60, 61, 65, 69
  synovium, 18, 37
Reiter's disease, 112
Remodeling, 21–25, 31, 87, 89, 92, 99, 128, 129
Renal (*see also* Uremia), disease, 30, 64, 89, 92, 139, 140, 153, 154
  excretion of urate, 137
  insufficiency, 27, 64, 94
  osteodystrophy, 33, 93, 94
  rickets, 89
  stones (*see* Nephrolithiasis)
Reticulin, 6, 10, 44
Rheumatic fever, 114, 120, 124, 126, 127, 155
Rheumatism, 114
Rheumatoid arthritis (*see also* Juvenile rheumatoid arthritis), 63, 72, 101, 111, 114–123, 125, 126, 142, 147, 155, 156
Rheumatoid factor, 116–119, 121, 125–127, 149, 155
Rheumatoid nodule, 120, 121, 124, 149
Rice body, 115
Rickets, 27, 28, 88–93, 152, 154
Rubella, 109, 110

Sacroiliac joint, 145, 147–149
Scheuermann's disease, 151
Schmorl node, 85, 151
Schwann cell, 49, 50, 67
Sciatica, 145, 150
Scleroderma, 63, 72, 124–126, 155
Sclerosis, bone, 95, 97, 99, 103, 104, 108
Scoliosis, 73, 79, 151
Scurvy, 4
Sequestrum, 106–108
SGOT, 49, 69, 152
Shoulder-hand syndrome, 84
Sialic acid, 8, 20
Sickle cell anemia, 95, 108, 142
Sinus tract, 106, 108, 111
Sjögren's syndrome, 71, 126, 127, 155
Sodium, 13, 14, 43
Somatomedin, 31, 77, 78, 79
Spondylitis (*see also* Ankylosing spondylitis), 105, 108
Spondylolisthesis, 151
Spondylosis, 150, 151
Staphylococcus, 105, 112
Still's disease (*see also* Juvenile rheumatoid arthritis), 121–123, 132
Strain, 2
  lumbosacral, 145
  tears, 151

Stress (see also Pressure), 2, 3, 20, 32, 100, 151
Subluxation, 115, 116, 119, 130, 147
Sudeck's atrophy, 83, 84
Synovial, fluid, 9, 10, 37, 39, 40, 111, 115, 118, 129, 139, 140, 144, 155, 156
  mucin, 39, 40, 111, 119
  osteochondromatosis, 39, 132, 134
Synovitis, 114–116, 126, 129
Synovium, 36, 37, 39, 110–115, 117, 119, 128, 132, 133, 135
  lining cells, 18, 39, 41, 115, 116, 118, 123
Syphilis, 108, 109, 124, 132, 149, 155
Syringomyelia, 132, 133
Systemic lupus erythematosus, 71, 72, 124, 125, 142, 155

Temporomandibular joint, 35, 73
  condylar cartilage, 73, 79
Tendinitis, 114, 115
  calcific, 140
Tendon, 1, 4, 20, 44, 47, 64, 112, 114, 148
  receptor, 51
Tenosynovitis, 112, 114, 115
Tetany, 33, 56, 58, 92, 94
Thalassemia, 95
Thyroid hormone, 31

Tidemark, 35, 36, 38
Tophus, 136, 138–140
Toxoplasma, 60
Trichinosis, 52, 60
Trypanosomiasis, 60, 117
Tuberculosis, 107–108, 111, 112

Urate, 133–140, 153, 154
Uremia, 14, 30, 33, 140

Vertebra (see also Codfish vertebra), 41, 73, 77, 97, 103, 105–109, 145, 151
  compression fracture, 85, 86, 89
Vitamin D, 28–30, 33, 88–90, 92, 94
  1,25-$(OH)_2D_3$, 28–30, 89
Volkmann canal, 25, 106
Von Kossa stain, 12, 88, 90

Water, 1, 2, 7, 8, 11–15, 20, 37, 41
Whipple's disease, 122
Wolff's law, 25, 31, 73, 83, 97
Woven bone, 11, 13, 21, 81, 104

Xanthine oxidase, 135–138

Yaws, 109